QUALITY
CONTROL
IN
HAEMATOLOGY

QUALITY
CONTROL
IN
HAEMATOLOGY

SYMPOSIUM OF THE INTERNATIONAL COMMITTEE
FOR STANDARDIZATION IN HAEMATOLOGY

edited by

S. M. LEWIS

Royal Postgraduate Medical School, London W12 0HS, England

and

J. F. COSTER

National Institute of Public Health, Bilthoven, The Netherlands

1975

ACADEMIC PRESS: London, New York, San Francisco

A Subsidiary of Harcourt Brace Jovanovich, Publishers

ACADEMIC PRESS INC. (LONDON) LTD
24/28 Oval Road,
London NW1 7DX

United States Edition published by
ACADEMIC PRESS INC.
111 Fifth Avenue
New York, New York 10003

Library of Congress Catalog Card Number: 75–19656
ISBN: 0–12–446850–0

Text set in 11/12 pt Monotype Baskerville, printed by letterpress,
and bound in Great Britain at The Pitman Press, Bath

CONTRIBUTORS

BRIAN S. BULL, M.D., *Professor and Chairman, Department of Pathology and Laboratory Medicine, Loma Linda University School of Medicine, Loma Linda, California 92354, U.S.A.*

I. CHANARIN, B.Sc., M.D., F.R.C.Path., *Head of Section of Haematology and Consultant Haematologist, Medical Research Council Clinical Research Centre, Northwick Park Hospital, Harrow, Middlesex HA1 3UJ, England.*

J. F. COSTER, M.D., *Chief Bacteriologist, District Laboratories, National Institute of Public Health, P.O. Box 1, Bilthoven, The Netherlands.*

P. J. CROSLAND-TAYLOR, M.D., F.R.C.Path., *Senior Lecturer, Department of Haematology, Middlesex Hospital, London W1P 7LD, England.*

RUSSELL J. EILERS, M.D., *Bio-Science Laboratories, 7600 Tyrone Avenue, Van Nuys, California 91405, U.S.A.*

ALAIN F. GOGUEL, M.D., *Professeur Agrégé, Faculté de Médecine, Paris Ouest—Hématologie, Hôpital Ambroise Paré, 921000 Boulogne, France.*

A. H. HOLTZ, Ph.D., *National Institute of Public Health, P.O. Box 1, Bilthoven, The Netherlands.*

G. I. C. INGRAM, M.D., F.R.C.P., *Professor of Experimental Haematology, St Thomas's Hospital Medical School, St Thomas's Hospital, London SE1 7EH, England.*

JOHN A. KOEPKE, M.D., *Professor and Vice-Chairman of Pathology Department, University of Iowa, Iowa City, Iowa 52242, U.S.A.*

S. M. LEWIS, B.Sc., M.D., F.R.C.Path., *Reader in Haematology, Royal Postgraduate Medical School and Consultant Haematologist, Hammersmith Hospital, London W12 0HS, England.*

W. LÖHR, M.D., *Anatomisches Institut der Albert-Ludwigs Universität, 78 Freiburg i. Br., Albertstrasse 17, West Germany.*

C. C. MERRY, M.B., *Assistant Professor of Pathology, University of Manitoba Faculty of Medicine, and Director of Section of Haematology, Health Sciences Centre, 700 William Avenue, Winnipeg, Manitoba R3E 0Z3, Canada.*

D. W. PENNER, M.D., *Associate Professor of Pathology, University of Manitoba Faculty of Medicine, and Director of Section of Haematology, Health Sciences Centre, 700 William Avenue, Winnipeg, Manitoba R3E 0Z3, Canada.*

J. E. PETTIT, M.D., M.R.C.Path., *Senior Lecturer, Department of Haematology, Royal Free Hospital Medical School, Hampstead, London NW3 2QG, England.*

L. POLLER, M.D., F.R.C.Path., *Consultant Haematologist, Withington Hospital, West Didsbury, Manchester 20, England.*

L. S. Sacker, M.D., F.R.C.Path., *Consultant Haematologist, St Stephen's Hospital, Fulham Road, London SW10 9TH, England.*

Robert M. Schmidt, M.D., M.P.H., *Director, Hematology Division, Center for Disease Control, 1600 Clifton Road, Atlanta, Georgia 30333, U.S.A.*

O. W. van Assendelft, M.D., *Laboratory of Chemical Physiology, Bloemsingel 10, Groningen, The Netherlands.*

P. G. Ward, L.I.Biol., L.R.I.C., *Scientific Officer, Haematology Quality Control Assessment Scheme, Department of Haematology, Royal Postgraduate Medical School, London W12 0HS, England.*

I. M. Weisbrot, M.D., *Northwalk Hospital, Connecticut 06856, U.S.A.*

D. Wittekind, M.D , *Anatomisches Institut der Albert-Ludwigs Universität, 78 Freiberg i. Br., Albertstrasse 17, West Germany.*

PREFACE

During the past few years, the International Committee for Standardization in Haematology (ICSH) has sponsored a series of symposia at congresses of the International Society of Haematology. The purpose of these symposia has been to review activities of ICSH and its Expert Panels in selected fields and to consider the scope for future work in order to pursue the constitutional aims of ICSH, namely, "to promote the development of international standards that are needed to achieve international comparability of results of haematology analysis; to promote and encourage the improvement of methods and standards; and to maintain a forum for communication amongst component organizations to promote improvement in haematology laboratory function".

Previous symposia have dealt with haemoglobinometry, documentation, standardization of haemocytometry and of packed cell volume determination, automation in haematology, haematology of normal man, haemoglobin structure and measurement, standardization of serum iron and iron-binding capacity assays, computers in haematology, iron metabolism, standardization in abnormal haemoglobins and thalassaemias. Some of the papers presented at these symposia have been published in regular numbers of journals, or as a special supplement to a journal or in book form. Academic Press has already published two such ICSH books.

The present book continues this tradition as it is based on the symposium which took place in Jerusalem in September 1974 at the 15th Congress of the International Society of Haematology. On this occasion, however, we have also included some chapters from colleagues who were unable to attend the symposium and who, we believed, would make valuable contributions which would ensure that the subject was dealt with as comprehensively as possible.

We would like to express our appreciation to Dr G. Izak who was the Organizing Secretary of the Congress, for placing all facilities at our disposal, to our colleagues who have contributed to this publication, and to the staff of Academic Press from whom we have received every possible help. Although we have accepted editorial responsibility for the publication of this book, it stems essentially from the activities of ICSH and we are grateful for the support and co-operation of our colleagues of the ICSH Board and Secretariat. Finally, our thanks are also due to the World Health Organization from whom we have received great encouragement as well as a grant which has provided some financial support for our work.

S. M. LEWIS

J. F. COSTER

London and Bilthoven
June 1975

FOREWORD

Quality control and standardization of laboratory methods are closely related subjects which have a direct bearing on the primary objective to be achieved in clinical and public health laboratories: to provide a reliable basis for an aetiological diagnosis and for a scientifically sound treatment of the patient or epidemiological approach to disease.

A Symposium on Quality Control was organized by the International Committee for Standardization in Haematology during the fifteenth Congress of the International Society of Haematology in Jerusalem in September 1974, and the work presented at that Symposium is recorded in this volume. Shortly thereafter, in November 1974, the Director General of the World Health Organization convened a consultation meeting in Geneva with experts of various laboratory disciplines on the standardization of diagnostic methods and materials.

It is essential that quality control be practised by every laboratory and this should include both intra-laboratory and inter-laboratory procedures. In some respects quality control programmes are similar in various disciplines and by carrying out a collaborative programme some duplication could be avoided. As far as inter-laboratory trials are concerned, certain countries carry out sophisticated schemes which require complex computer programs, mass production of test samples, speedy transit of material and reports. Quality control schemes in these countries also include provision of reference preparations and calibration standards, control of commercial reference preparations by certification, and evaluation of instruments and diagnostic kits. These activities can profitably be co-ordinated by the World Health Organization and expanded to other countries in a co-ordinated scheme. An associated agreed scheme for certification and instrument evaluation could also avoid duplication of national efforts for the necessary controls in the rapidly expanding market of instruments and reagents.

At a less sophisticated level, quality control requires the availability of reference preparations and knowledge of simple statistical methods, in order to ensure an adequate level of precision. Laboratory personnel should be trained in these procedures as part of their general training. Here again international co-operation by bilateral and multilateral approach is of great importance.

Haematology is a discipline that serves a wide range of specialities in clinical and preventive medicine and covers a wide variety of techniques:

analytical biochemistry, physical measurement, cell morphology and stain technology. Each of these techniques requires its proper system for quality control and standardization, based on scientifically sound principles and practically acceptable methods.

The editors and the authors of this publication must be commended for their initiative and continuous efforts and for their valuable contributions towards achieving this objective.

DR J. SPAANDER
Chairman
International Committee for
Standardization in Haematology

National Institute of Public Health
Bilthoven
The Netherlands
June 1975

CONTENTS

1. Principles of Total Quality Control

RUSSELL J. EILERS

University of Kansas Medical Centre,
*Kansas City, Kansas, U.S.A.**

I. Concept of Total Quality Control

A. INTRODUCTION

In this era of consumerism, one of the greatest challenges to medicine is quality assurance in health care provided to the patient public. For the director of a medical laboratory this must mean a total quality control system that assures the provision of high quality laboratory results at a reasonable cost. The experiences of application of quality control engineering in other industries has evolved the principles of quality control.† In the past twenty years some of these principles have gradually been applied in the medical laboratory, by individual laboratorians, professional associations or governments. What is needed is a system of total quality control for the medical laboratory.

Total quality control can be defined as an effective system for integrating the quality development, maintenance, and improvement effects of various individuals within a laboratory so as to enable production of services at the most economical level which allow for the full satisfaction of the physician and patient (Feigenbaum, 1961).

In the phrase "quality control", quality does not have to mean in the

---

* Bio-Science Laboratories, 7600 Tyrone Avenue, Van Nuys, California 91405, U.S.A.
† This chapter is adapted for the medical laboratory from principles of total quality control outlined in the textbook by Feigenbaum (1961).

popular sense "best", but could mean best for certain patient conditions. These conditions must take account of the actual use to which the laboratory results will be put by the physician to solve the individual patient care problems and also the cost of the procedures to the patient. The word control represents four aspects of management activity: (1) setting standards for quality; (2) appraising conformance to these standards; (3) acting when the standards are not met; (4) planning for improvements in the standards.

Total quality control is an aid, not a substitute, for action by the physician director in the proper selection of a test for a particular care problem, the selection of proper method and instrumentation to be used in the laboratory, and conscientious inspection of the activities within the laboratory. Total quality control must start at the top with the director but must be practised by the entire laboratory staff. Whilst a medical laboratory may require the mass production of laboratory results, these are individual results for individual problems of individual patients. Accordingly, the total quality control programme must be designed to control the overall process as well as the individual result. The laboratory's motto must be "to *prevent* poor quality rather than to *correct* poor quality". Thus, the quality control programme must look at the entire process beginning in the office or at the bedside with the selection of proper test for a particular patient care problem and collection of the sample to the output of a quality result for reliable interpretation at the bedside by the physician to help resolve the patient care problem.

Most quality control methods are old, but what is new is the integration of these unco-ordinated activities within the medical laboratory into an overall administrative programme. As they are developed, new quality control techniques must be incorporated that may be useful in dealing with, thinking about, or increasing the emphasis on the reali-ability of the laboratories' results by improving their accuracy, precision, sensitivity or specificity in obtaining that result.

B. DEFINITION OF TERMS TO CHARACTERIZE METHODS

Accuracy is the degree of agreement of individual or average specimens with an accepted reference value or level (ASTM E-456-72), or the quantitative expression for the deviation of a spectro-chemical determination from an accepted reference level (ASTM E-135-72a). The reference value should, of course, be the true value as far as this can be ascertained (see p. 14).

Precision is the degree of mutual agreement among individuals measurements. Relative to a method of test, precision is the degree of mutual agreement among individual measurements made under prescribed like conditions. The imprecision of measurements may be characterized as the standard deviation of errors of measurement (ASTM E-456-72).

Sensitivity (detection limit preferred) is a stated limit value that designates the lowest concentration or mass that can be estimated or determined with confidence and that is specific to the analytical procedure used (ASTM E-135-72a).

Specificity is the ability of a method of assay to determine only the highly pure constituent in contrast to other components that may be present in the matrix of the sample.

C. BENEFITS

Utilization of a total quality control programme may provide the following benefits for a laboratory: (1) improvement of the quality of individual laboratory results; (2) reduction in the operating costs and losses of a laboratory; (3) improvement in employee morale; (4) discovery of reduction of problem bottle necks in the processing of laboratory procedures in the laboratory: (5) improved inspection methods for the director; (6) sounder setting of productivity standards for personnel; (7) schedules for preventive maintenance of equipment; (8) provision of a data base for the laboratory director to assure himself and the consumer physician of the reliability of the laboratory result for excellence in patient care; (9) provide a factual basis for cost accounting of criteria to disregard or reprocess a sample and the cost involved in the inspection of results.

D. THE ELEMENTS AND JOBS OF QUALITY CONTROL

The elements or factors of quality control can be divided into two broad categories: (1) technological: the methods and instruments used, raw materials, the process and control of process; (2) human: the attitudes, knowledge and skills of the laboratory staff—clerks, aides, technicians, technologists, supervisors and doctoral-level laboratory personnel.

More has been written about the technological factors, but the most important are the human factors. These factors can be grouped into the jobs of quality control. For purposes of the medical laboratory, the jobs will be divided into six phases.

II. THE SIX PHASES OF TOTAL QUALITY CONTROL

A. PHASE I—DESIGN CONTROL

This means a proper laboratory facility design and a pattern of staffing to ensure efficient handling of the anticipated workload of the selected assay systems for the range of health care problems to be served by that laboratory. Will the laboratory just serve inpatients, outpatients or be a regional reference laboratory? Will the laboratory just be providing services for the care of acute and chronic medical diseases or will it also be involved in the screening of normal and near-well patients? What percentage of the workload is anticipated to be emergency work? The facility design must consider the functions within the laboratory and the flow of work from units it has interrelationships with (Thomas, 1974).

The proper selection of methods and instruments for assay systems will depend upon the basic science education and professional training in various disciplines of the laboratory staff. This will also depend on their participation in continuing education and clinical research activities concerning the use of laboratory methods via their professional societies and medical institutions.

The addition of new laboratory procedures, especially screening procedures for outpatients and inpatients, should be done with full consultation of the medical staff. Screening procedures should allow immediate follow up by the laboratory staff on abnormal results with other available tests that may aid in elucidating the abnormal result and point to a specific diagnosis. This should reduce the patient's hospital stay, decrease laboratory costs and expedite general health care service. The laboratory staff must learn to eliminate old, insensitive, inaccurate or imprecise procedures from the list of available laboratory tests and to replace these procedures with others that will provide rapid, meaningful results that aid the differential diagnosis or treatment.

The laboratory should have an organizational chart. Duties and responsibilities for each position in the chart should be defined. Task and function analysis studies will aid in selecting the category of laboratory worker for each of these positions.

B. PHASE II—INCOMING MATERIAL CONTROLS

Incoming material control involves receiving or stocking, at the most economical levels of quality, only those materials and equipment whose quality conforms to the laboratory's specification requirements. A laboratory may meet this requirement by purchasing (a) reagent grade chemicals as defined by the American Chemical Society (b) Standard Reference Materials for highly pure chemicals as available from a National Bureau of Standards or International Standards Organization, (c) standards produced by an industry but certified by a professional organization, or (d) products that are labelled as being produced according to specifications of an authoritative body. Batches of quality control materials must be selected at the levels needed to control the process. Some batches of material may be purchased as unassayed material, usually at a more economical cost, whereas others will have to be purchased as assayed by an accepted reference method in order to calibrate as well as to control the process of automated instruments. Certified physical standards can be purchased from a National Bureau of Standards to check the calibration of absorbance and wave length scales of instruments or other essential characteristics of laboratory equipment (temperature, weights). A mechanism must be developed or a published evaluation mechanism utilized for the evaluation of reagent kits or systems (see p. 193). Products can also be purchased which have been evaluated by a professional organization or a national government body. All volumetric glassware including pipettes should be checked for calibration before being put into routine use.

Many countries now have national committees for clinical laboratory standards which aid in the definition of specifications for reagents, methods, instruments and modes of practice that can be utilized. Ideally, these committees should be a federation of all interested parties such as professions, government and industry working together to develop and accept standards by a consensus voting mechanism. On the international level there are diverse international standardizing bodies that promulgate standards for all areas of the clinical laboratory. WHO has indicated willingness to co-ordinate the standardization activities of international organization for the clinical laboratory, especially for the control of *in vitro* diagnostic kits and reagents (WHO, 1974).

A laboratory should have a laboratory manual that defines the specifications for the specimens to be utilized in the process of constituent assays. This should recommend the proper collection equipment and anticoagulants, the stability time of constituents if processing is delayed and how to ship if it must go over a distance. If drugs may interfere with the process, information should be requested on drugs the patient is on. The instructions should recommend the time of day to draw the sample, especially in relation to meal-time; and should also specify whether the patient may be sitting, standing or must be lying prone, as posture of the patient has a significant effect on the haemoglobin, PCV and other quantitative results, and must be considered when evaluating the assay results (see p. 214).

Since quality work begins with qualified personnel, the human elements of laboratory staff need to be controlled as a raw material. Job descriptions should be available for each category of laboratory worker utilized. These should define education and professional requirements and laboratory experience necessary. For each position in the organizational chart there should be available some description of duties and responsibilities. This will assist the interviewer in evaluation of the prospective employee's background and attitudes for the job.

C. PHASE III—PROCESS CONTROL

Process control involves an internal programme for control of the assay procedures which includes preventive maintenance of the equipment, an external programme of interlaboratory trials to evaluate the proficiency of the laboratory during process and/or a regional study group to help refine the precision and accuracy of the assay procedures.

The laboratory procedure manual should detail, for each constituent assay, the procedure for use of a known highly pure standard material in a known solvent to calibrate the method and the use of control materials at appropriate levels in each batch of determinations for a particular constituent to assure that the overall process is in control. Large batches of stable control materials should be utilized so that a laboratory can determine the limits for within-day variation of the process and compare this to the variation for between-days as well as months of the year to document the precision of the assay procedure and its stability over time. A regional group of laboratories could utilize the same pool of control materials so that they can compare the precision of their laboratories for a particular method and/or instrumentation for determining the same constituent. Statistical analysis should be performed by a sponsoring agency that allows the laboratory director to document the internal quality control efforts of the laboratory, to observe the trends of precision of the methods within the laboratory over time, and to compare the laboratory's precision with that of peers utilizing the same measurement system and a selected reference field method. Periodically, there should be included unknown control materials that have been assayed by an accepted field reference method or absolute base method as determined by a national authoritative body, in order that the bias of the laboratory's methods can be defined according to these reference points. Thus, a regional study programme can serve the needs of an internal quality control programme as well as an external programme to refine the precision and accuracy of the assay procedures.

For external quality control the laboratories receive unknown samples of the expected range of levels of constituents and report their results to a sponsoring agency for statistical analysis (interlaboratory trials). From the agency the laboratory should receive for each constituent assayed a report on the laboratory precision, the accuracy of their results compared to other laboratories using the same methodology and instrumentation, the bias of its measuring system compared to an accepted reference method and a statement of acceptability of their results. Thus, a mechanism is established to monitor the quality of the work of laboratories involved.

Preventive maintenance is an important quality control technique. It requires a regularly scheduled examination of all equipment involved in the assay procedure before failure occurs, and thereby assures correct functioning (Hamlin *et al.*, 1974). The methods of checking and testing must be defined in the laboratory procedure manual for each piece of equipment. Adequate records of documentation should be kept on their performance as well as a record of repair of instruments. Trends may be defined where certain components need periodic replacement and thus records can document the intervals of down-time for complicated pieces of equipment and their probable causes.

Each member of the laboratory staff should have a proper attitude to quality control and have defined responsibilities relating to the quality control programme of that laboratory. These should be consistent with the job description of personnel category. An educational programme within the laboratory and review sessions should be held to consider the effectiveness of the quality control programme of that laboratory and the quality control jobs carried out by each member while performing his duties at the bench. A documentation scheme should be established where unknown samples are linked to specific personnel to define a record of their proficiency and skill in operating certain pieces of equipment and carrying out various assay procedures. This should include replicate precision of each technologist by comparison with most competent technologist, and duplicate counts or results of duplicate blind samples. If proficiency testing indicates deficiencies, there should be refresher training or re-education of the personnel involved to correct such deficiencies. As new assay procedures are introduced into the laboratory a documentation system should be introduced to ensure that the personnel are checked out on the equip-ment and are proficient with the new procedure. Physical examinations of personnel should be carried out at intervals to determine the presence of colour blindness or the need for a physical aid, such as glasses, or to detect physical conditions caused by hazards that they may be exposed to. There should be adequate documentation of accidents that occur on the job and this should be correlated to the specific personnel file.

D. PHASE IV—OUTPUT CONTROL

Output control in Phase IV must assure the provision of quality laboratory reports in a meaningful format for the physician. The laboratory reports should provide identification of patient, his location and physician, the date and time of collection of sample, gross description and source of specimen when pertinent. A consistent terminology, format and usage of abbreviations and symbols should be used. It is high recommended that the IFCC recommendations for SI metric system of quantity and units be utilized (Dybkaer and Jorgensen, 1967). The laboratory report should be easy to prepare and provide for presentation of results into groups for physiological significance, specific organ function or specific disease evaluation, in order to facilitate evaluation by the physician. Exceptionally high and low values should be rechecked for proper processing and to ensure that calculations are correct before they are entered on the report. Significant results should be flagged to bring them to the attention of the physician at the bedside. There should be clear delineation of abnormal results from normal results. This means the provision of a normal value reference system by the laboratory for their assay procedures. Normal values must be determined by that laboratory and not be dependent upon values published in literature especially if the methods utilized are modified. The laboratory report sheet should be in a logical and accessible location in the medical chart and have administrative and record-keeping value (Burns *et al.*, 1974).

E. PHASE V—RELIABILITY CONTROL

Reliability control should ensure that utilization of assay results correlates with health care needs. Does the laboratory result help or confuse the physician in solving the patient care problem for which it was ordered? The laboratory should not just provide results of qualitative and quantitative tests, but should provide, with these results, meaningful normal values with confidence limits that relate to the precision, accuracy, sensitivity and specificity of the assay methods.

The laboratory staff should utilize a variety of external communications to meet this obligation. A laboratory manual should be made available to the nursing staff and physicians outlining the proper collection of patient specimens, methods to be utilized within the laboratory, and expected normal values. Subsequent laboratory bulletins should be utilized to report revisions of the methods or new laboratory procedures being offered and their relationship to normal and pathophysiology. Laboratory staff should make rounds to discuss the problems technologists are having within the laboratory with patient specimens and the assay requests or methods, to discuss the problems of the nursing staff with the collection of patient specimens or obtaining laboratory results and the physician should be contacted directly concerning supposed laboratory error problems. The patient's physician should also be contacted directly whenever an unusual result is found that should be brought immediately to his attention.

Laboratory reports are only a one way communication mechanism. The laboratory staff must try to ensure that the results are received, the vagaries of specimen collection, accuracy and precision of methods are understood and the proper response ellicited from the physician. Laboratory staff are expected to be involved in the interpretation of cell patterns in the blood smears and the urinary sediment and the patterns of protein electrophoresis. The laboratory staff should become involved in and aid in the interpretation of results of batteries of laboratory procedures requested for health maintenance screening and the results of batteries of procedures done on admission or repeated during the hospital stay. For this obligation, the laboratory staff will have to develop a laboratory data analysis system.

Criteria for health information systems require that the system must be problem oriented, persons specific, yet population based and parsimonious. (White et al., 1972). Tools are available to help meet these criteria. The laboratory staff should work with the medical staff to establish and use the problem oriented medical chart (Weed, 1971). This recording method defines the individual patient care problems from the patient data base (history, physical examination, and basic laboratory data) and documents further laboratory, clinical, or therapeutic procedures utilized for each problem and the effect of these procedures on the patient problems to provide clinical conclusions of diagnosis and patient outcome. To facilitate these correlations of the diverse problems and procedures to clinical conclusions, a systemized nomenclature of medical language such as that being developed from SNOP is essential (Wells, 1973). Both of these tools will support the four phase process of multivariant laboratory data analysis system needed to synthesize and interpret multiple laboratory results to aid the physician (Grams et al., 1972). The four processes handle analytic error limits, multivariant normality evaluation, multivariant diagnosis pattern recognition, and trend analysis of assay results over time for their effect on diagnosis or therapy. This avoids the pitfalls of over-utilization of procedures when only univariant analysis of multiple laboratory tests is attempted. All significant results are brought to the attention of the physician.

F. PHASE VI—VERIFICATION CONTROL

Verification control involves investigative studies to locate the causes of defective results and to determine the possibility of improving the quality characteristics of results. Some investigations have been referred to in the Phase III external programmes of interlaboratory trials and regional study groups and the internal proficiency testing programme for personnel.

The ultimate overall investigative study is an inspection and accreditation programme of the 'medical laboratory sponsored by a professional organization e.g., the Inspection and Accreditation Program of the College of American Pathologists or the American Association of Blood Banks. In each programme, broad standards are defined as to proper operation of the medical laboratory. Detailed check lists are utilized during the inspection process to verify and document the overall total quality control system of that medical laboratory. The general standards and checklists cover personnel, facilities, methods and instrumentation, preventive maintenance programme, internal and external quality control programmes, record keeping and personnel safety.

Another form of verification control is a workload recording system that documents the productivity of the medical laboratory. Laboratories must use a uniform thesaurus and a system for counting raw laboratory procedures done within the laboratory to allow comparison to other laboratories in a country or the world. A realistic mechanism must be used to transform the raw count of procedures carried out into productivity figures of the technical staff which reflect the influence of semi-automated or automated devices. The question to be answered is whether proposed changes in methodologies or technology improve the productivity of technical staff. Does it reduce the laboratory cost to the patient or to the third party paying for the service? In North America the Canadian Association of Pathologists and the College of American Pathologists have published a workload recording system that utilizes time engineered weight factors to convert raw count to productivity figures for the technical staff (Laboratory Management and Planning Committee, 1972). A time engineering study plan has been developed to define what a weight load factor should be for a new method or technology. This allows laboratorians to develop weight load factors and share them with their colleagues around the world. The use of this workload recording system allows the laboratory director to predict technical staff and space needs as indicated by the workload figures. The system allows the separation of workload relating to the output of patient sample results from that involved in the quality control of process. Thus, there is the beginning of a cost analysis system for the medical laboratory.

III. Cost of Quality Control

Although costs are associated with a total quality control system, the system provides a means of measuring and optimizing the cost of quality control activities (Feigenbaum, 1961). The cost of quality control can be distributed for analysis into four broad categories. (1) Prevention costs: expenses for keeping defective results from occurring by having a planned quality control programme and the expenses involved in the development of such a programme. (2) Appraisal costs: the expenses for operating and maintaining an internal quality control programme, participating in external quality control programmes and an inspection and accreditation programme. (3) Internal failure costs: the expense for redoing or scrapping a batch of specimens or an individual specimen due to some element of improper processing that leads to an improper result. (4) External figure costs: the investigation of all complaints of the physician, or patient consumer or failure of the laboratory result to help solve the patient care problem.

A medical laboratory should invest most of their funds in the prevention and appraisal cost to minimize the expenses of repeating batches or individual specimens or investigating the complaints of physicians about supposed laboratory error. This would be consistent with the laboratory's motto "to prevent poor quality rather than to correct poor quality" (see chapter 17).

A total cost analysis system requires a charter of income and expense accounts, an organizational chart with job descriptions for each defined position, the utilization of a workload recording system and a task and function analysis mechanism to assure the proper mix of laboratory personnel to be utilized for the selected assay systems to provide the workload of results at the most economical cost.

IV. ROLE OF STATISTICS

An understanding of statistics has become a requirement for laboratorians to help cope with the problems of a clinical laboratory (Barnett, 1971). Although statistics are useful in the overall quality control programme, statistics are only a part of the total quality control plan and should not become the plan itself. Four statistical tools of common use in quality control activities are: (1) frequency distributions; (2) control charts; (3) sampling tables; (4) special methods such as test of significance, correlation techniques and analysis of variance. Application of statistical tools are discussed in chapters 2, 5 and 9.

REFERENCES

ASTM E-135-72a. *In* "1973 Annual Book of ASTM Standards", Part 30, pp. 274–280. American Society for Testing and Materials, Philadelphia.
ASTM E-456-72. *In* "1973 Annual Book of ASTM Standards", Part 30, p. 1413. American Society for Testing and Materials, Philadelphia.

Barnett, R. N. (1971). "Clinical Laboratory Statistics". Little, Brown and Co., Boston.

Burns, E. L., Hanson, D. J., Schoen, J., Barnett, R. N., Minekler, T. and Winter, S. (1974). *Am. J. clin. Path.* **61,** 900–903.

Dybkaer, R. and Jorgensen, K. (1967). "Quantities and Units in Clinical Chemistry". Williams and Wilkins Company, Baltimore.

Feigenbaum A. V. (1961). "Total Quality Control, Engineering and Management". McGraw-Hill Book Company, Inc., New York.

Grams, R. R., Johnson, E. A. and Benson, E. S. (1972). *Am. J. clin. Path.* **58,** 177–219.

Hamlin, W. G., Duckworth, J. K., Gilmer, P. R., Jr and Stevins, M. V. (1974). "Laboratory Instrument Maintenance and Function Verification". College of American Pathologists, Chicago.

Laboratory Management and Planning Committee (1972). "A Workload Recording Method for Clinical Laboratories", 2nd edn. College of American Pathologists, Chicago.

Thomas, R. G. ed. (1974). "Manual for Laboratory Planning and Design". College of American Pathologists, Chicago.

Weed, L. W. (1971). *Archs intern. Med.* **127,** 101–105.

Wells, A. H. (1973). *Med. Rec. News* **44,** (2), 12–16.

White, K. L., Mornaghon, B. A. and Gaus, C. R. (1972). *New Engl. J. Med,* **287,** 1223–1227.

WHO, (1974). "Standardization of Diagnostic Material. Report of the Director General to Twenth-seventh World Health Assembly". A 27/12, April 17, 1974. World Health Organization, Geneva.

2. Concepts of Inter-laboratory Trials: An International Haematological Survey

O. W. VAN ASSENDELFT

Laboratory of Chemical Physiology, University of Groningen, The Netherlands

and

A. H. HOLTZ

National Institute of Public Health, Bilthoven, The Netherlands

I. INTRODUCTION

Clinical laboratories, including the haematological, are faced with an increasing workload as well as an increasing complexity and an ever broadening range of tests. At the same time there is an increased demand for a greater validity of the test results. It has been recognized that this demand can only be met adequately when every factor in the analytical process is appropriately controlled. These factors include the training of personnel, the performance of instruments, the quality of reagents and reference preparations and, finally, the way in which tests are actually performed. The term *quality control* usually refers only to the control of the last mentioned factor. It involves assessing, by means of control samples, the accuracy and the precision of the methods being applied in daily practice. Because statistical techniques are used in this assessment, the procedure is sometimes also called *statistical quality control*. It is in this limited sense that the term quality control will henceforth be used in this chapter.

Two facets can be discerned in quality control: (1) internal or intra-laboratory control—analysis of control samples selected daily by the laboratory head, in order to check precision and accuracy within the laboratory; (2) external or inter-laboratory control—analysis of control samples received periodically from an outside source, in order to compare accuracy levels of different laboratories. A procedure is said to be under control when both the accuracy and the precision of the results are within generally accepted, though arbitrary, limits. Accuracy is a measure of the closeness of an estimated value to its true value. It is defined by reference to an absolute standard, whenever possible based on international specifications, through secondary standards which are normally used in the laboratory. Precision is a measure of the reproducibility of an estimation. It is assessed by repeated measurements of a control preparation.

The practice of internal quality control originated spontaneously, its need being self-evident. The demand for external quality control did not arise, however, until it became apparent that differences, sometimes large, existed between results in different laboratories. The wish to obtain comparable results is motivated not only from dissatisfaction from the analytical point of view, but also by such practical reasons as the frequent movement of patients from one health service institution to another. This stresses the importance, medically as well as economically, of the necessity of comparing new laboratory data with previous ones. Furthermore there exists an increasing need for a wider range of reference values ("normal values"), i.e. separate values valid for subjects of differing age and sex, and under differing conditions. As yet few reliable data in this field are available. It will be necessary for a large number of laboratories to co-operate if this comprehensive work is to be carried out effectively. An inter-laboratory control scheme is of the utmost importance for the reliability of values acquired for this purpose.

The importance that society in general accords to these principles is illustrated by the fact that legislative measures have been or are being drawn up, whereby certification of laboratories or authorization of payment for tests can be made dependent upon proficiency testing results. Inter-laboratory control can offer valuable documentation on the quality of work being done.

This chapter deals with requirements of inter-laboratory controlling as well as with problems which have been encountered in the field of haematology, especially as the international level.

II. AIMS AND ORGANIZATION

Inter-laboratory trials should be seen as a contribution to the ultimate goal of laboratory standardization; i.e. obtaining test results with systematic errors approaching zero and random errors as small as possible. The immediate objective of a trial can either be *quality control*, i.e. controlling of the participating laboratories, or *quality study*, i.e. study of the performance of tests or procedures. Both purposes may be served in trials organized at a local, a regional, a national or an international level.

In the case of quality *control*, the primary purpose of inter-laboratory trials is to provide an independent check on the accuracy of the participating laboratories, it being up to the participants themselves to use the results to rectify possible deficiencies in their internal quality control programmes. Quality *study* will stem from scientific curiosity as to either the performance of a certain test or procedure, or of a certain group of laboratories.

At the local, regional or national level, inter-laboratory trials are organized by the heads of the participating laboratories as a group, through the services of a professional society or of a foundation specifically created to this end. The programmes benefit the laboratories themselves as well as the patients serviced by them. The necessity for these projects has usually arisen spontaneously, as was the case in internal quality control. In some cases the necessity arises through outside pressure, e.g. because of certain legislative measures.

At the international level inter-laboratory trials are organized by international professional societies or committees, usually as part of larger projects, sometimes at the invitation of, or on the instruction of, international governing bodies, e.g. Council of Europe, Commission of the European Communities, Eurotransplant, International Atomic Energy Agency, World Health Organization.

III. PREPARATION FOR A TRIAL

When planning an inter-laboratory trial, decisions must be made on the following points: the aim; the component or test; the set-up;

participants; the sample; material; instructions and result sheet; packaging and despatch.

A. AIM

The aim of the trial should be clear from the very beginning because the objective sought, quality control or quality study, determines the set-up. A trial, for instance, in which the participants are asked to perform a determination only once under routine conditions cannot be called the ideal frame for conclusions as to the intrinsic value of certain procedures in comparison to others. The spread of results will be caused by both systematic errors due to method differences and random errors due to daily variability. Minimizing the role of random errors through multiplicity of determinations causes systematic errors to show up more clearly.

B. COMPONENT OR TEST

The component(s) to be determined or the test(s) to be performed may be chosen for a variety of reasons. It may be desirable to investigate the reliability of the determination of a component, whatever the method used, or the reliability of one or more methods for the determination of the same component. The component may be an important, widely determined one or it may be less widely determined though equally important.

On the other hand, it may be the intention of the trial to find optimal conditions within a method (method standardization), or to check on an established standardized method when it is applied in practice.

In general, the test(s) should be performable and the component(s) be measurable under routine laboratory conditions, not only in laboratories having sophisticated equipment, but also in those without such facilities.

C. SET-UP

The set-up of a trial is determined by its aim. When the objective is quality control, trials are usually held at regular intervals varying from a few weeks to a few months. It is customary to ask the participants to treat the testing material as a patient sample and to report only a single value obtained under routine conditions; no more material is sent than is ordinarily required for one such determination. When the objective of the trial is a quality study, the participants are asked to do repeated determinations for a predetermined number of days. In addition, reference solutions and/or samples of material other than the original sample (e.g. blood haemolysate comparable with whole blood) may be included for analysis, in order to obtain information on intermediate steps of the analytical procedure.

Quality control trials may be carried out with or without reference laboratories. Reference laboratories must meet certain requirements such as have, for instance, been formulated by the Medical Society of West Germany. They should be independent, have a comprehensive system of internal quality control, continually participate in comparative studies with other reference laboratories, and have the facilities to apply basic analytical procedures and to compare different methods.

The results obtained by reference laboratories are used to judge the values obtained by the participants. In those cases where reference laboratories are not included, it is assumed that the mean value found by the participants represents, or is close to, the true value. Theoretically this is not correct, e.g. methods with systematic errors may have been used; in practice, however, this approximation proves acceptable in many cases.

The reference laboratories may analyse the samples at the time of the trial or in advance. Analysis at the time of the trial has certain advantages, e.g. when the component being determined is not absolutely stable with time. Analysis beforehand may also be advantageous, e.g. providing certainty that the samples being sent meet the necessary requirements.

D. PARTICIPANTS

In quality control trials the aim is to include all laboratories within the designated area; in quality study trials the number of participating laboratories will have to be restricted. The choice of participants may be based on size, capacity or geographical distribution of laboratories on the one hand, or on expertise or the function of the laboratory head on the other hand. Care, however, should be taken to make as representative a selection as possible, especially when international bodies or local contact persons have been asked for address lists.

E. SAMPLE

In particular in the case of quality control the sample should, in general appearance, resemble a patient sample as closely as possible, e.g. as regards colour and viscosity. The amount will be dictated by the aim of the trial. In quality control trials there should be sufficient, but not more than the amount necessary for a duplicate determination; in quality study trials, the amount must be sufficient for the number of determinations required.

The number of samples is dictated by the concentration range to be tested and by the type of statistical analysis to be made. The component concentration in the sample(s) depends on their number. When only one sample is sent a component concentration on the borderline normal/abnormal is considered optimal for quality control purposes. When more than one sample is sent, concentrations should be chosen which include one which is considered to be normal and others which occur under pathological conditions. Whenever possible, preference should be given to concentrations falling well within the normal range, the abnormal low and the abnormal high ranges.

F. MATERIAL

With a view to the unavoidable timelag between the preparation of the samples and their analysis by the participating laboratories, as well as the often uncontrollable conditions during mailing, the material should not deteriorate with time or on violent movement, nor be influenced by temperature changes. For these reasons it may be necessary to freeze-dry the material. The component must be stable for a certain, known, period of time as well as within a certain, known, temperature range.

Conserving or bactericidal agents may only be added if they do not interfere in any way with the determination. If freeze-dried material is used, the possibility of errors made on reconstitution should be seriously considered. These errors may be due to impure solvents, to erroneous dilution or to instability of certain components immediately after reconstitution (e.g. alkaline phosphatase). Finally care must be taken to ensure that the material is free of Australia antigen for all practical purposes.

G. INSTRUCTIONS AND RESULT SHEET

Directions for participating laboratories should be given on a separate sheet. They should include a short description of the material and instructions as regards handling of the sample. The latter must be detailed and include the conditions of storage, the reconstitution of the sample if necessary, the homogenization, time and number of determinations and conditions of analysis (i.e. whether to treat as a patient sample or with special precautions). Finally, instructions must be given on how the result sheet is to be completed, and it has proven advantageous to include, as an example, a mock filled-in result sheet.

The result sheet should allow for data on the following: name and address of sender; date of arrival of samples; condition of samples on arrival; date of analysis; value(s) obtained; method of analysis, with attention to possible modifications; method of calibration; reagent(s) used, with attention to their preparation; instrument(s) used; date of despatch of samples (to be filled in by organizer); deadline for returning results (to be filled in by organizer).

As regards the method(s), reagent(s) and instrument(s), use is often made of a list of possibilities on which the laboratory has only to check-off the method, reagent and instrument it is using. Although this method is convenient for both participant and organizer, the result is often unreliable inasmuch as some people believe they are using a certain method or reagent, while on further investigation this proves not to be the case. Reliable information can be obtained by having the participants describe the method and reagent composition themselves. Sorting these data, however, is time consuming for the organizer and can cause delay in reporting the trial results.

As regards reporting of values obtained, it is advisable to prescribe the number of decimal places because a varying number may hamper statistical analysis.

In large trials, when there are many participating laboratories, it may be worthwhile to make use of precoded result sheets or cards allowing for mechanical or electronic handling. In all cases the sheets should be available in duplicate, e.g. carbon copy, the participants keeping a copy on file.

There will be a tendency not to return complicated or extensive sheets, unless the trial is on a compulsory basis. Therefore one must not overdo the exercise.

H. PACKAGING AND DESPATCH

The size of the bottle containing the sample should be adapted to the amount, taking into account the possible necessity of homogenization. The neck of the bottle must be of sufficient size to allow pipettes to pass through. Capping must not only be absolutely leak-proof but the whole container must meet rather strict requirements such as those which have been formulated for containers holding patient material in general. Specifications drafted by the International Standards Organization (ISO) are worth studying in this context. Care must also be given to the packaging of samples, taking into account their possible exposure to violent handling on transport.

Existing postal and custom laws must allow uninhibited movement from place to place, state to state, or country to country. Any regulations concerning packaging and despatch of medical samples should be consulted. In order to speed up arrival at the participating laboratories, it is wise to label the material "No commercial value. For scientific purposes only."

Participants should be advised in advance that trial samples will be arriving shortly. If possible the samples should arrive at such a time that immediate processing is possible. Local holidays, vacation periods and week-ends should be taken into account.

IV. Evaluation of Results

After a predetermined time, not much after the deadline set for the return of answers, the results received are prepared for presentation and statistical analysis. In particular in quality control programmes a trial is of value to the participants only when the end result is available within a comparatively short time. In a quality study this time limit may be appreciably longer, although participants will appreciate a preliminary summary of the results to compare their values with those obtained by the other laboratories. Although tedious, it often proves worthwhile to look through the result sheets and correct obvious mistakes before the results are processed (see also chapter 5).

A. PRESENTATION OF DATA

Depending on the set-up of the trial the results may be grouped as to method first. The final comprehensive survey may be given as tables or graphs (histograms). Both methods have advantages and disadvantages. Tables are easy to prepare, also by a computer; they are, however, less illustrative. Graphs are better in this respect; their preparation, however, is time consuming. It should be noted that a kind of frequency-diagram can also be produced by a computer (see chapter 3).

B. STATISTICAL ANALYSIS

It is of major importance to realize the value and especially the limitation of statistical procedures, when applied to characterize a set of numbers (a population sample) such as is obtained in a trial.

The simplest procedures are the determination of the median or of the mode, together with the range, or the calculation of mean, together with the range and/or the standard deviation (s.d.). The median, defined as the point on the scale having equal numbers of observations above and below it, is of little or no value in the given case. The mode, defined as the number occurring most frequently, may be an acceptable parameter to indicate the centre of a population when there is a non-Gaussian distribution. The range indicates the spread of the group of numbers.

Care should be taken not to confuse the mode with the mean. The mean (arithmetic mean) or average, defined as the numerical mean value calculated from a series of numbers, is of significance only when there is a Gaussian distribution, i.e. when the values are not biassed by systematic errors. The standard deviation is defined as a measure of the dispersion of a group of numbers around their arithmetic mean.

With all these statistical manipulations one is not confronted with the actual error (i.e. systematic plus random error) unless the true values are available through the inclusion of reference laboratories in the trial. For this reason the technique introduced by Youden is being applied more and more in interlaboratory trials.

C. YOUDEN PLOTS

It has become usual in trials to have the participating laboratories analyse (a multiple of) two samples simultaneously for the same constituent, and to display the results obtained on each pair of samples in a graph. The values found for sample A are plotted along the abscissa, those for sample B along the ordinate. A cluster of points is thus obtained. This technique was devised by the statistician Youden, and the graph is consequently called a Youden plot.

In the original design the two samples were required to be similar in composition and quite close together as regards content. For each level of the content range under study, a separate pair of samples needed to be analysed. In recent years it has become customary to combine samples of differing content within one plot, e.g. low and normal, or normal and high. The requirement of similarity in composition has, understandably, remained unchanged. Although more instructive, this modification has complicated the construction of the Youden plot (see below).

In addition to being a concise and convenient way of presentation, the Youden plot can give an indication of the type of errors which have been made. To this end lines demarcating the allowable ranges for the results obtained on the two samples are drawn in the plot (Fig. 1). These ranges may be based on mean and s.d. of the results themselves, or on mean and s.d. of the results obtained by the reference laboratories. Plus or minus twice the value of the s.d. is often taken as the allowable range. Further, a diagonal is drawn in the central area (c) extending into quadrants 2 and 4. All results falling within the central area are acceptable, as regards both systematic and random error. Results lying outside the central area in quadrants 1 and 3 are due to random errors, most of those in quadrants 2 and 4 (namely those around the diagonal line) are due to systematic errors. The "area of systematic errors" is often demarcated by lines which are extensions of the diagonals of the upper left and lower right quadrant of the central area. Drawing these lines in this way is, however, a simplification (see below).

When the contents of two samples differ, the Youden plot can be made to appear more regular by adaptation of the scales to each other, e.g. by compressing the one or expanding the other, so that the allowable ranges for sample A and sample B cover equal distances on the axes (Fig. 2). The same can be obtained by taking the respective standard deviations as unit distances on the axes. The central area now becomes

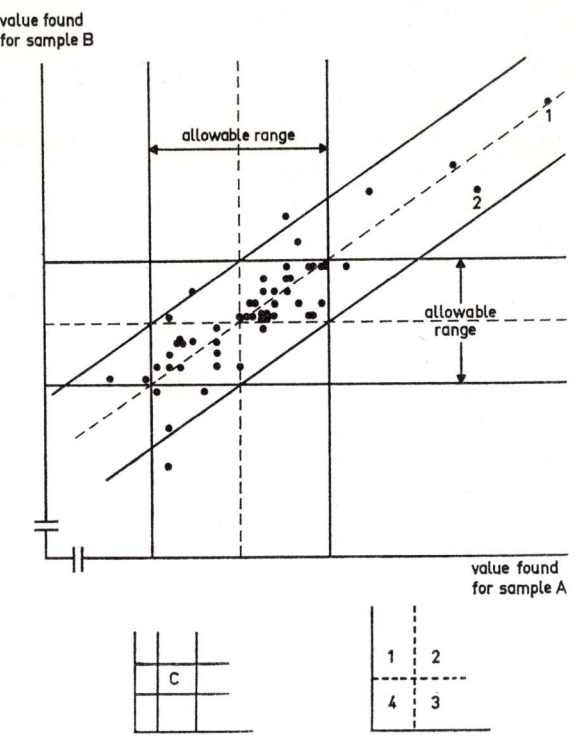

Fig. 1. Example of a Youden plot. Scale on abscissa equal to scale on ordinate. (Explanation given in the text.) Difference in allowable range is due to larger standard deviation of values for sample A as compared to sample B, because of concentration differences. The values marked 1 and 2 in the graph refer to the same values marked 1 and 2 in Fig. 3. Lower left: C = central area. Lower right: 1, 2, 3 and 4 = first, second, third and fourth quadrant.

a square. It can, however, better be given by the theoretically more correct form of a circle. The slope of the area of systematic errors has become 45° (i.e. the direction of the diagonal of the square), and the tangents to the circle are now taken as lines demarcating this area, which results in a shifting of the limits for acceptable results.

When constructing a Youden plot, the following points should be kept in mind.

(1) Positioning the area of systematic errors at a slope determined by

the ratio of the standard deviations for samples A and B (as has been done in Fig. 1) makes sense only when this slope equals, or at least differs only slightly from the slope of the line of symmetry of the cluster of points. The bisector of the two possible regression lines may be taken as line of symmetry of the cluster of points. If this is not the case,

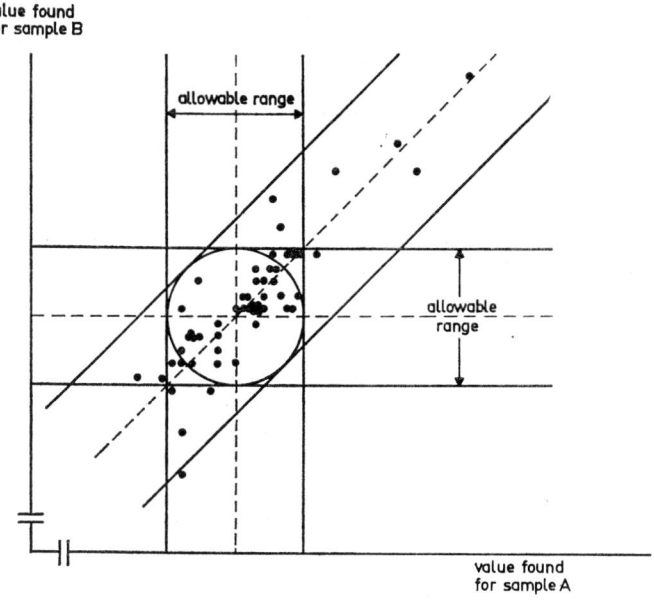

FIG. 2. Youden plot of Fig. 1, after compression of abscissa and slight distention of ordinate. Explanation given in the text.

the area of systematic errors will be rotated with respect to the cluster of points (Fig. 3). In consequence erroneous conclusions may be drawn concerning results found to lie outside the limits, e.g. the error in results 1 and 2 (Fig. 3) seems to be due to a systematic and random component while in fact only a systematic component is present (Fig. 1).

Differing orientation of the two areas can occur when standard deviations are used other than those calculated from the actual results, e.g. from those of the reference laboratories. However justified these other standard deviations may be, they should be handled with caution in judging deviating results.

(2) Positioning the area of systematic errors at 45° with respect to the axes, is correct only when the standard deviations for sample A and sample B are equal, or have been made to cover equal distances in the graph. The situation in which the s.d. for sample A was equal to

that for sample B was of course the case in Youden's original design, where the contents of the two samples were (practically) the same. It is, however, seldom the case when samples of differing content are used, the s.d. of the results most likely being proportional to the content.

(3) Finally, care should be taken not to give the unwary reader a false impression by cutting off part of the axes or by using illogically differing scales on the axes.

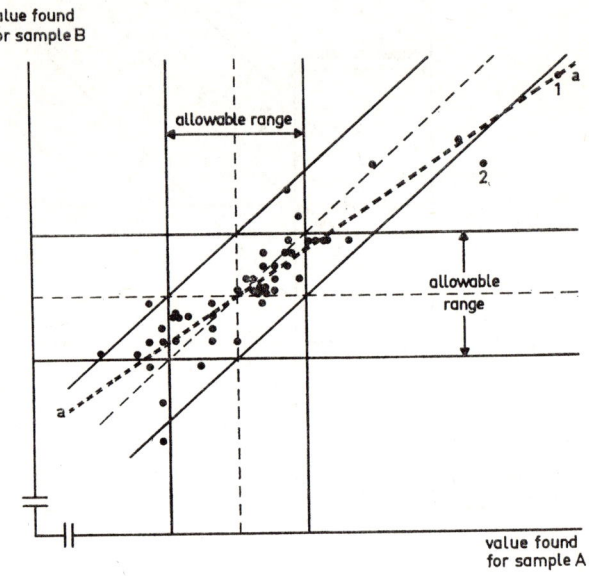

Fig. 3. Youden plot of results depicted in Fig. 1 with arbitrary allowable ranges drawn in; a–a = line of symmetry. Explanation given in the text. The values marked 1 and 2 in the graph refer to the same values marked 1 and 2 in Fig. 1.

V. FEEDBACK OF INFORMATION

The report to the participants of a trial must include a comprehensive survey of all values found and a commentary giving information on how the results have been analysed and what could have been the cause of deviating results. The survey of the results may be in the form of tables or of graphs (compare presentation of data). It should be noted that differentiation of the results as regards methods used and/or instrumentation is always useful in order to show systematic differences, if any. Care must be taken not to try to present too much information within too little space, a defect to which tables are particularly prone. Readability of the report is of prime importance and graphs are usually distinctly advantageous in this context. It is useful to keep in mind that frequency diagrams and even Youden plots can be produced as computer print-out.

In order to optimize the usefulness of a quality control trial report, each participant should receive a personal copy on which his values may be identified at a glance. This may be attained by special mention of his values or by marking them directly in the tables or graphs. One may not take for granted that the participants will have at hand results they have obtained previously. It is obvious that because of this the trial organizer will have to exercise extra care to keep the report as a whole anonymous.

Having to make personal copies of the trial report does not, however, exempt the organizer from the obligation of having the report ready within as short a time as possible. Many laboratories will use trial results, especially if the trial is held regularly, to judge the efficiency of their internal control scheme. Because deviating results may endanger a laboratory's certification, timely reporting of results is definitely indicated in the case of compulsory trials in which some scoring system of laboratories is incorporated. If, for some reason or another, it is inevitable that the complete report cannot be sent within a few weeks of the date of sample analysis, preliminary information, e.g. the "true value" of the samples only, should be sent to the participants.

In some cases, especially in quality control trials, follow-up may be indicated. The trial may result in certain methods proving inadequate or it might be concluded that other reference solutions are necessary. Sometimes the material used as testing samples in the trial is made available afterwards as reference material to the participants. This enables the laboratories to repeat the determinations they have made for the trial and to adjust their method, if necessary. Thus, for example, in a recent haemoglobinometry trial involving laboratories which had received, in the near past, the International Haemiglobincyanide Reference Preparation, proof was obtained for the additional need of a blood-like preparation to control the method in daily practice (see below).

VI. An International Haemoglobinometry Trial

Of the many tests performed in the haematological laboratory, few lend themselves as yet for inter-laboratory trials on an international scale. Until now, international experience has really been obtained by The International Committee for Standardization in Hematology (ICSH) only with the determination of the haemoglobin content of blood, although some data are available also on the determination of the erythrocyte count and packed cell volume (PCV) value of whole blood samples (Holtz, 1970).

To illustrate some of the points made in the preceding section, a summary of a recent survey of the haemoglobin determination in different countries, is given. This survey was organized in 1973 by the National Institute of Public Health in the Netherlands at the request of the Expert Panel on Haemoglobinometry of ICSH. The objective was to obtain information on the state of haemoglobinometry, in particular in those laboratories which had applied for the International Haemiglobincyanide (HiCN) Reference Preparation (ICSH, 1972a, b).

The invitation to participate was sent to some 95 people, primarily recipients of the HiCN preparation in Africa and Europe. Fifty four HiCN recipients participated (46 in Europe, eight in Africa) as well as a further 13 persons associated with ICSH, bringing the total participation to 67.

The testing material consisted of two fresh blood samples, two glycerol-containing haemolysates, and a haemiglobincyanide solution. The blood contained EDTA (disodium salt), 1·4 g/l, as anticoagulant. The haemolysates had been prepared according to a method used at the Center for Disease Control, Atlanta, Ga, U.S.A. The HiCN solution was identical to the material used as International Reference Preparation. For the blood samples and the haemolysates, the haemoglobin concentration had to be determined; in the case of the HiCN solution the optical density at $\lambda = 540$ nm.

In general the samples reached their destination within 3–4 days after despatch. This interval was short enough to prevent detrimental changes, even in the whole blood samples.

The "correct" values of the samples were derived from the results obtained by the members of the ICSH Expert Panel on Haemoglobinometry. Their method was that set out in the ICSH recommendations (ICSH, 1967). The values obtained are shown in Table I.

TABLE I. Values of samples derived from results of the ICSH Expert Panel on Haemoglobinometry

Sample	Value
1 (whole blood)	17·2 g/dl
2 (whole blood)	13·8 ,,
3 (haemolysate)	15·6 ,,
4 (haemolysate)	12·5 ,,
5 (HiCN solution; measured in a 1·00 cm square cuvette)	$D^{540} = 0.405$

The results of the haemoglobin determination of the participating laboratories are summarized in Figs 4–7, grouped as to the type of measuring instrument used. As expected, the HiCN method had been used in all laboratories, although with differently composed reagents and different dilution ratios. The reagent according to Van Kampen-Zijlstra or a similar solution was most commonly used (ferricyanide,

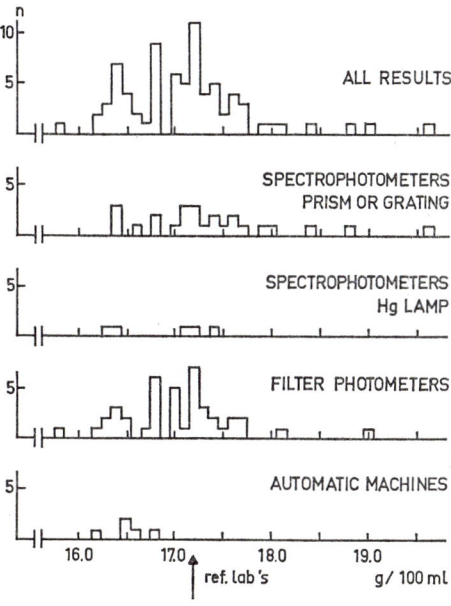

FIG. 4. Trial results sample 1, whole blood. Grouped as to measuring instruments. Reference laboratory value indicated by arrow.

cyanide, phosphate and a detergent); other reagents used were the Van Kampen-Zijlstra solution without detergent, the original Drabkin reagent (ferricyanide, cyanide), the modified Drabkin (ferricyanide, cyanide, bicarbonate) and, in the case of automatic machines, specific reagents, sometimes of unknown composition. The dilution ratios varied between 1/100 and 1/500, 1/250 being most common.

Because the majority of participants could be expected to have calibrated their method using HiCN reference solutions, it was disappointing to find, even after excluding some outliers, that the range of the results was rather large, up to $\pm 1 \cdot 5$ g/dl. These figures are similar to the results obtained in the trial of 1968 (Holtz, 1970). There is little difference between the results obtained on the blood samples and those obtained on the haemolysates, although the distribution curves seem to be more "normal" for the latter. It is further remarkable that

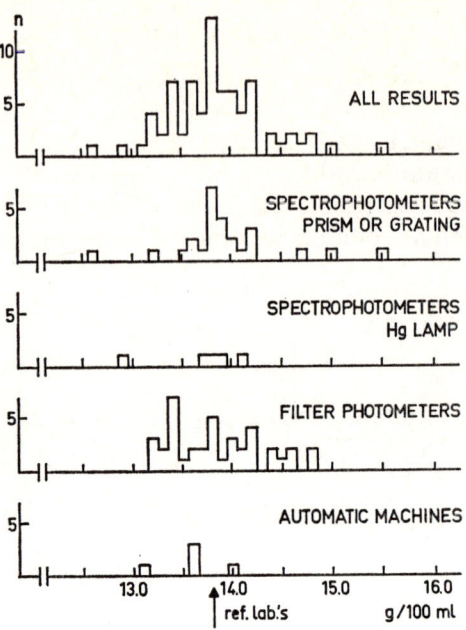

Fig. 5. Trial results sample 2, whole blood. Grouped as to measuring instruments. Reference laboratory value indicated by arrow.

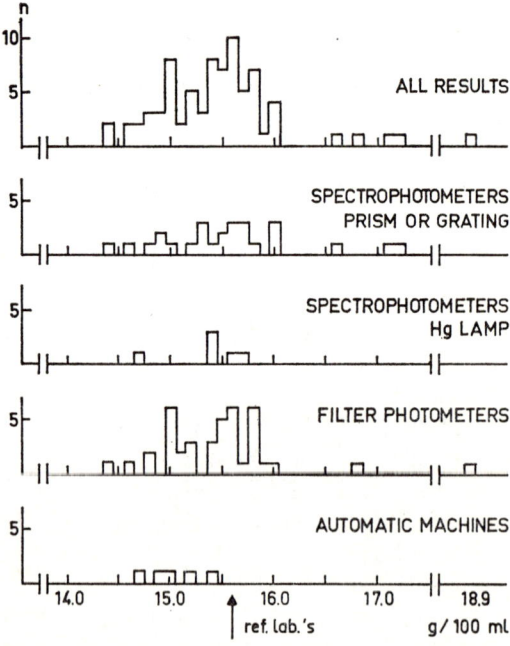

Fig. 6. Trial results sample 3, haemolysate. Grouped as to measuring instruments. Reference laboratory value indicated by arrow.

laboratories where a spectrophotometer was used as measuring instrument did not perform better than those using filter photometers. The results of the automatic machines (four Coulter S' and one Technicon SM 12) seem to be consistently lower than the general mean.

In Figs 8 and 9 the results of the two blood samples and of the two haemolysates have been combined, after subdividing as regards

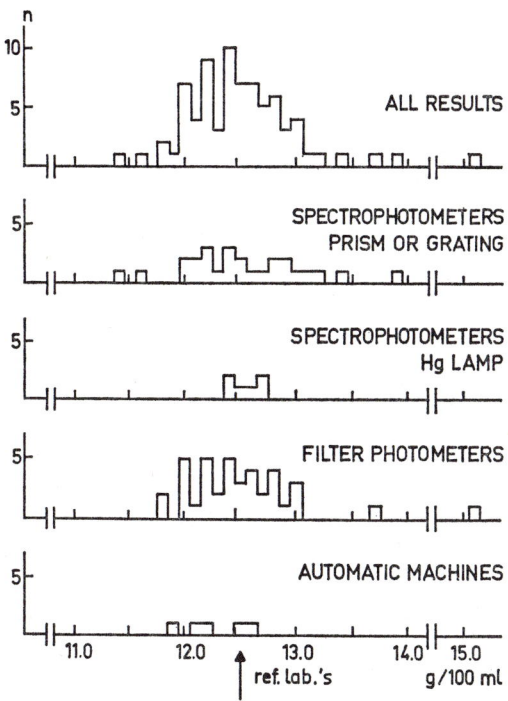

FIG. 7. Trial results sample 4, haemolysate. Grouped as to measuring instruments. Reference laboratory value indicated by arrow.

methods of dilution, in Youden plots (Skendzel and Youden, 1969). It is clear from the elongation of the cluster of points along the diagonal that many systematic errors have been made. Where, in the haemoglobin determination procedure, this is the case, will be discussed further on.

The results of the optical density measurements of sample 5, the HiCN solution, again grouped as to the type of measuring instrument, are summarized in Fig. 10. The values found with spectrophotometers (Bausch and Lomb, Beckman, Cary, Jouan, Optica, Shimadzu, Unicam, Zeiss PMQ) are fairly close together (0·392–0·408; one outlier),

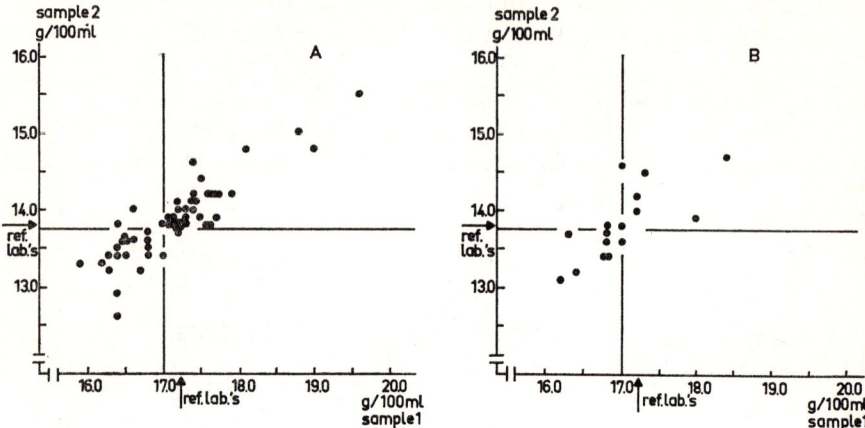

Fig. 8. Youden plots of samples 1 and 2, whole blood. Reference laboratory values indicated by arrows. Solid line denotes mean value of all results. A, dilution made with pipettes; B, dilution made with a diluter.

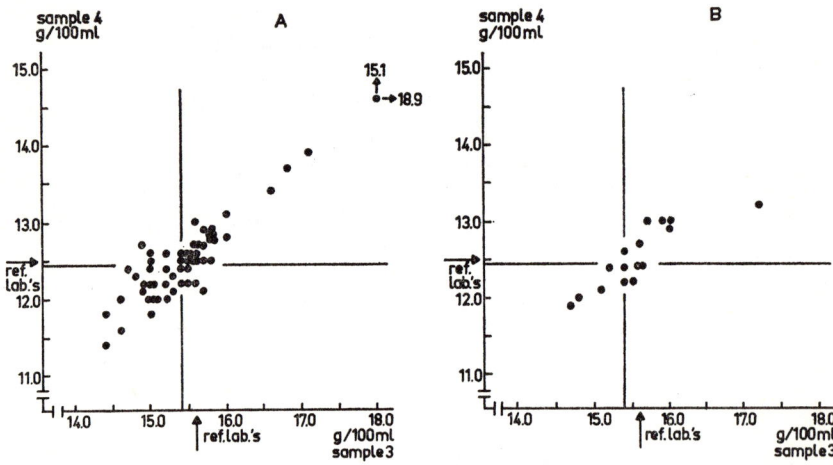

Fig. 9. Youden plots of samples 3 and 4, haemolysate. Reference laboratory values indicated by arrows. Solid line denotes mean value of all results. A, dilution made with pipettes; B, dilution made with a diluter.

except for those of the Jena Spekol which show a much wider range and are consistently lower (0·334–0·386). An explanation for this phenomenon cannot be given, as the Spekol, supposedly a spectrophotometer, should not give differing readings. As could be expected, the results of the filter photometers are completely arbitrary, ranging from 0·278 to

0·410. Since the values had been obtained in 1·00 cm square cuvettes or, where not, had been corrected to this lightpath length before they were noted in the graphs, it must be the quality of the filter i.e. its band width, which determines the level of the optical density value.

In Fig. 11, the optical density values of the HiCN solution have been converted into corresponding haemoglobin values of a hypothetical

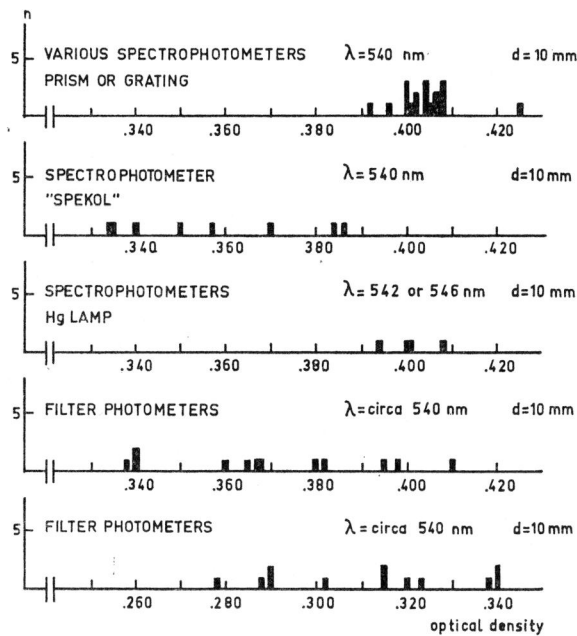

Fig. 10. Trial results sample 5, HiCN solution. Grouped as to measuring instruments. Reference laboratory value $D_{\text{HiCN}}^{540} = 0\cdot405$.

blood sample, using for each laboratory its own conversion factor (deducible from the measurements of the blood samples and the haemo-lysates), and a common dilution ratio. It is immediately evident that the values thus obtained have a much smaller range than those of the blood samples and the haemolysates (Figs 4–7). In the case of the Spekol instrument an individual calibration factor was calculated from the blood sample and haemolysate measurements. It should be noted that these instruments now perform equally well in comparison with the other spectrophotometers. It must, however, be concluded that optical density readings on these instruments do not reflect true optical density measurements.

For purposes of better comparison, the curves for "all results", as given in Figs 4–7 and 11 have been put underneath each other in Fig. 12. In Figs 13 and 14 differentiation is made between the various groups of instruments. It appears that the effect of the better precision is greatest for the spectrophotometers and hardly noticeable in the case of the filter photometers. In Fig. 14 the results obtained with Spekol spectrophotometers have been specially marked. Here it again appears

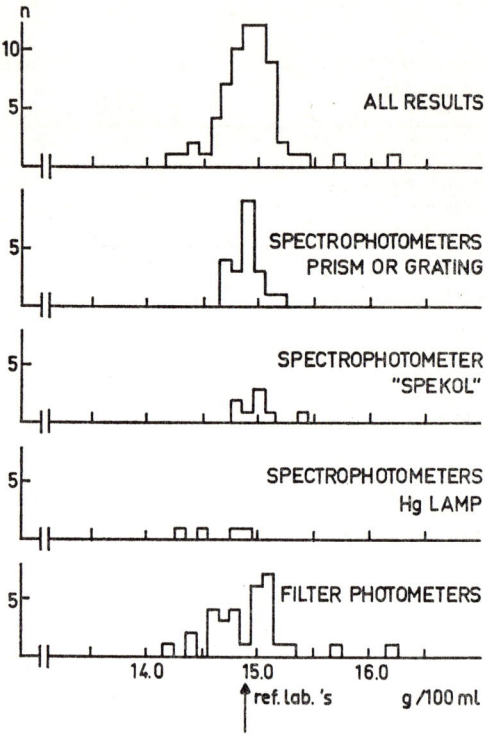

Fig. 11. Trial results sample 5, HiCN solution, converted to corresponding haemoglobin values taking a dilution factor 1:251. Results grouped as to measuring instruments. Reference laboratory value indicated by arrow.

that there is no difference in the performance of these instruments and spectrophotometers of other makes, provided the readings are used only relatively.

It is, of course, tempting to try to say which errors contribute most to the unsatisfactorily large spread of the haemoglobin values of the blood samples and the haemolysates. That they are of a systematic character has already been demonstrated (Figs 8 and 9). One should thus look for steps in the procedure where this type of error can be made. The following may be considered: (1) homogenization of the blood;

(2) pipetting of blood and reagent, or mechanically diluting; (3) conversion of Hb → HiCN; (4) optical density, as given by the photometer; (5) calibration factor, or calibrating of the photometer in Hb concentration.

(1) *Homogenization.* This can be ruled out because the spread for

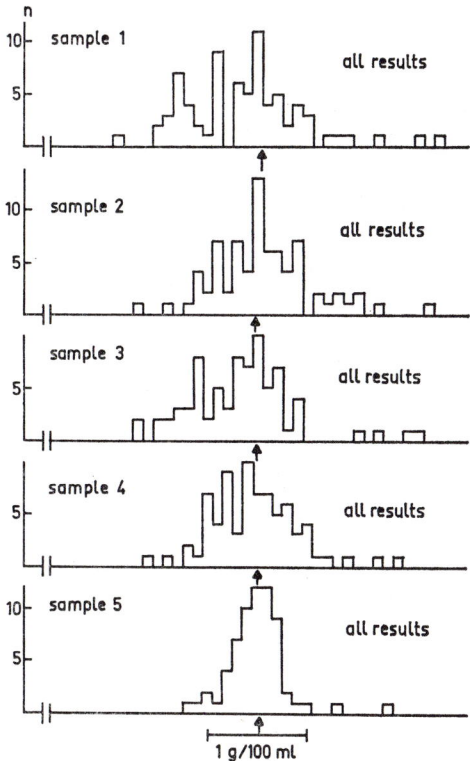

FIG. 12. Trial results of the five samples. Reference laboratory values indicated by arrows.

the whole blood samples and the haemolysates is practically equal (Figs 4–7).

(2) *Pipetting and diluting.* It is quite possible to make grave errors in this step. It has been shown more than once that blood pipettes are often calibrated wrongly (Lewis, 1970) and diluters may well be adjusted incorrectly. The strongest influence is, of course, to be expected from diluters and dispensers. In the present study, however, this did not prove to be the case. For the laboratories which had used pipettes only (45), the magnitude of the systematic error was the same as for

those which had used mechanical apparatus (diluter 19, pipette and dispenser 3), as is apparent from their respective positions in the Youden plots (Figs 8 and 9).

(3) *Conversion to HiCN*. It is, in principle, possible to make errors by waiting too short a time, and thus obtaining results with erroneously high values. It is not very likely, however, because the Van Kampen-Zijlstra reagent, which most people used, has a conversion time of less than 3 min.

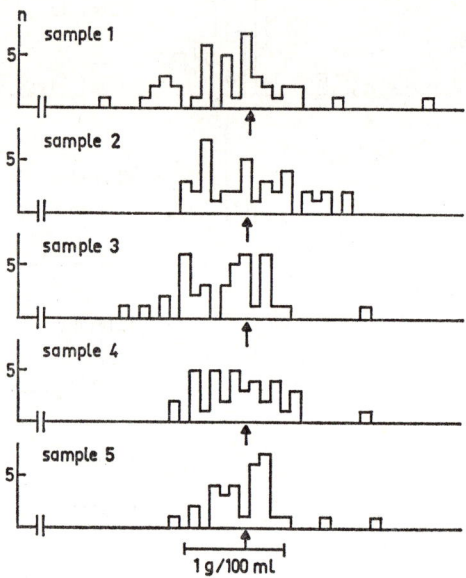

FIG. 13. Trial results, measuring on filter photometers, of the five samples. Reference laboratory values indicated by arrows.

(4) *Optical density measurements*. This is only applicable for spectrophotometers if the reading is used to calculate the haemoglobin concentration directly. It can be ruled out in this trial, because most results were obtained with filter photometers, and moreover, the optical density measurements of the spectrophotometers were quite good (Fig. 10). In the case of filter photometers, the reading is used only relatively, and, if the instrument gives a linear response, an error cannot be made.

(5) *Calibration*. Errors due to faulty calibration are possible in principle, but are certainly not solely responsible for the total spread in this trial since, if all steps prior to the optical density measurement are omitted—as has been the case for the ready-to-measure HiCN solution —the spread becomes smaller. In Fig. 12 the coefficient of variation for

the HiCN sample is 1·7%, for the whole blood and for haemolysate samples it varies between 2·9 and 3·3%.

The findings of this haemoglobinometry trial can be summarized as follows: the errors in the haemoglobin determination are mainly of a systematic nature, they are made in the dilution step of the procedure, as well as in the calibration. It is clear that, if diluters and dispensers are

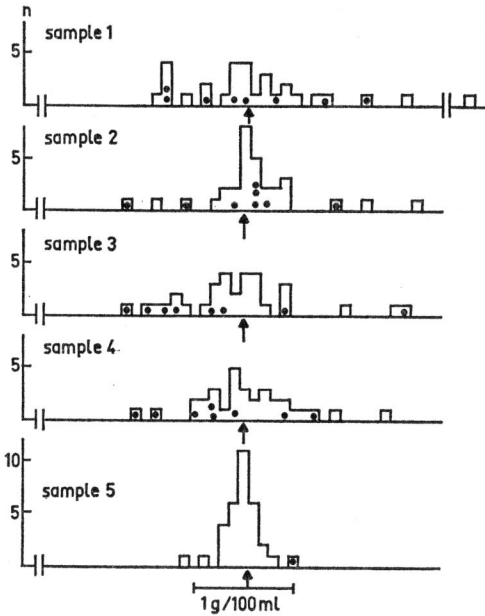

FIG. 14. Trial results, measuring on spectrophotometers, of the five samples. Values obtained using Spekol instruments indicated by dots. Reference laboratory values indicated by arrows.

used, systematic errors may be made. Since the majority of the laboratories had diluted manually, pipettes must also have been incorrectly handled (e.g. during filling with the sample, wiping off the excess, etc.). That incorrect calibration gives rise to systematic errors is clear without further explanation. Quantitatively, the calibration error (1·7%) is estimated to account for about one half of the total error (up to 3·3%), the dilution error (the difference of the total and the calibration error) for the other half.

It may be concluded from the results that, in addition to a HiCN reference solution to calibrate the instrument, a whole blood-like preparation is also necessary to control the method for determining haemoglobin in daily practice.

REFERENCES

Holtz, A. H. (1970). *In* "Standardization in Hematology" (G. Astaldi, C. Sirtori and G. Vanzetti, eds) pp. 113–120. Fondazione Carlo Erba, Milan.

International Committee for Standardization in Hematology (1967). *Br. J. Haemat.* **13** (Suppl.), 71–75.

International Committee for Standardization in Hematology (1972a). *Br. J. Haemat.* **23**, 123–124.

International Committee for Standardization in Hematology (1972b). *Lancet* **1**, 970.

Lewis, S. M. (1970). *In* "Standardization in Hematology" (G. Astaldi, C. Sirtori and G. Vanzetti, eds) pp. 45–50. Fonazione Carlo Erba, Milan.

Skendzel, L. P. and Youden, W. J. (1969). *Am. J. clin. Path.* **51**, 161–165.

BIBLIOGRAPHY

A detailed description of the determination of haemoglobin in blood, and of the errors which can be made, is given in the following publications.

Assendelft, O. W. van, Holtz, A. H. and Lewis, S. M. (1971). "Determination of the Haemoglobin Content of Blood". Document of the World Health Organization: WHO/BS/71.1026, WHO/HLS/71.46.

Assendelft, O. W. van, Zijlstra, W. G. and Kampen, E. J. van (1970). *Proc. K. ned. Akad. Wet.* **C73**, 104.

The following publications are concerned with general aspects of inter-laboratory trials and statistical methods.

Barnett, R. N. (1971). "Clinical Laboratory Statistics". Little, Brown and Company, Boston.

De Verdier, C-H. and Aronsson, T. (1970). *Scand. J. clin. lab. Invest.* **25**, 209–212.

Eilers, R. J. (1969). *5th med. J.*, Nashville, **62**, 1362–1365.

Stamm, D. (1971). *Schweiz. med. Wschr.* **101**, 429–437.

Strömme, J. H. and Eldjarn, L. (1970). *Scand. J. clin. lab. Invest.* **25**, 213–222.

Various authors. (1968). *Z. analyt. Chem.* **243**, 751–825.

Various authors. (1974). *Mitt. dt. Ges. Klinische Chem.* **2**, 25–56.

Von Boroviczény, K-G. and Merten, R. (1972). "Systematik der Qualitätskontrolle im Medizinischen Laboratorium". Medicus Verlag GmbH., Berlin.

Whitehead, T. P., Browning, D. M. and Gregory, A. (1973). *J. clin. Path.* **26**, 435–445.

Youden, W. J. (1969). "Statistical Techniques for Collaborative Tests". The Association of Official Analytical Chemists, Washington, D.C.

3. Inter-Laboratory Trials:
A National Proficiency Assessment Scheme in Britain

P. G. WARD and S. M. LEWIS*

Royal Postgraduate Medical School, London W12 0HS, England

I. Participation in Trials

In Britain, inter-laboratory proficiency trials have been organized since 1968 (Lewis and Burgess, 1969), under the direction of the British Committee for Standards in Haematology and more recently with financial support from the Government Department of Health (DHSS). There are now 350 participants. At monthly intervals the participants receive two specimens for blood counts (see chapter 6) and, also, appropriate material for one or other special investigation such as serum B_{12}, iron, abnormal haemoglobin.

Participants register according to the methods which they use (Fig. 1). In a trial, 10 ml volumes of each specimen are sent to users of fully-automated equipment (F) and 5 ml volumes to laboratories using manual (M) or semi-automated (S) methods.

The amount of data returned by a laboratory depends on whether the tests are performed by one or more than one method (Fig. 2). Data on specimens which are reported free of clots, lysis and/or contamination

* Supported by British Department of Health and Social Security.

IQC Form 11

BRITISH COMMITTEE FOR STANDARDS IN HAEMATOLOGY

INTERLABORATORY QUALITY CONTROL TRIALS: FULL BLOOD COUNT

INSTRUMENTATION RECORD

Please use a SEPARATE FORM for Manual, Semi-automated and Fully Automated Procedures.

NB: Please do NOT make any entry in open squares: except here

Please return to:
Dr S M Lewis
Department of Haematology
Royal Postgraduate Medical School
Hammersmith Hospital
London W12 OHS

Participant's Reference No — 1–3

Haemoglobin — `0` `1` 4–5

Procedure: ø Manual `M` Semi automated `S` Fully automated `F` 6

Comment

Instrument — 7–9

ø Filter `R` Grating `G` Prism `P` 10

Filter/Wavelength — 11–14

*Diluent — 15–17

Dilution — 18–21

*Standard — 22–24

Method of measurement: ø Calculation `C` Graph `H` Table `T` 25

Red Blood Cells — `0` `2` 26–27

Procedure — ø `M` `S` `F` 28

Instrument — 29–31

+ Aperture (Electronic) — 32–35

+ Attenuation (Gain) — 36–39

+ Threshold — 40–43

*Diluent — 44–46

*Lysing Agent —

Dilution — 47–50

White Blood Cells — `1` `0` 26–27

— ø `M` `S` `F` 28

— 29–31

— 32–35

— 36–39

— 40–43

— 44–46

— 47–50

Packed Cell Volume — `0` `3` 51–52

Procedure — ø `M` `S` `F` 53

Instrumentation/calculation — 54–56

+ Speed — 57–60

+ Time — 61–64

+ Reader — 65–66

Mean Cell Volume — `0` `4` 67–68

ø Instrumentation `I` Calculation `C` 69

Mean Cell Haemoglobin — `0` `5` 70–71

ø Instrumentation `I` Calculation `C` 72

Mean Cell Haemoglobin Concentration — `0` `6` 73–74

ø Instrumentation `I` Calculation `C` 75

Platelets — `0` `7` 51–52

Procedure — ø `M` `S` `F` 53

Instrument — 54–56

+ Aperture (Electronic) — 57–60

+ Attenuation/Gain — 61–64

+ Threshold — 65–68

*Diluent — 69–71

Dilution — 72–75

Date —

Signature —

ø Delete as appropriate
+ With fully automated equipment insert appropriate voltage potentiometer setting
* Detail composition/reference/manufacturer under 'Comment'

FIG. 1. Form for recording details of instrumentation and techniques used by participants in inter-laboratory trials.

are graded as satisfactory in one of four grades depending on the time taken for them to reach the participants. Grade "A" categorizes samples which reach their addressee by the day after posting. As a rule, 80% of specimens qualify as Grade A, and only results from these are used as the basis of all trial assessments. Thirty per cent of the results arrive back within 3 working days and 40% take 4 or 5 days. Returns are not received from one in five participants because of "pressure of work, equipment failure and/or the absence of the consignee". Most laboratories fail to report one or two trials in a year.

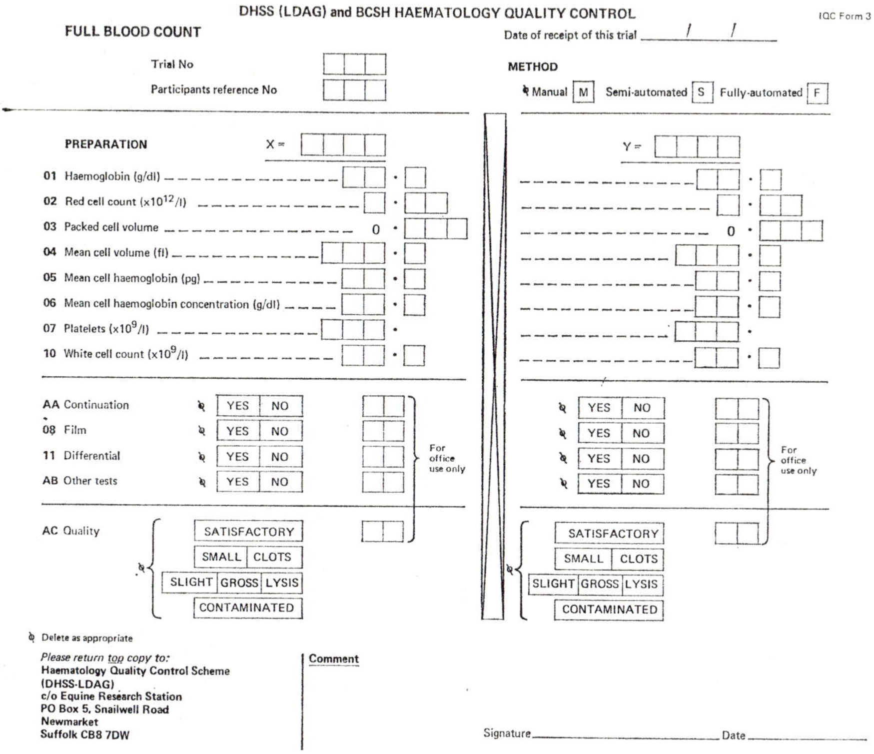

DHSS (LDAG) and BCSH HAEMATOLOGY QUALITY CONTROL

IQC Form 3

FULL BLOOD COUNT

Date of receipt of this trial ____ / / ____

Trial No

Participants reference No

METHOD

Manual M Semi-automated S Fully-automated F

PREPARATION X = Y =

01 Haemoglobin (g/dl)

02 Red cell count ($\times 10^{12}$/l)

03 Packed cell volume 0 ·

04 Mean cell volume (fl)

05 Mean cell haemoglobin (pg)

06 Mean cell haemoglobin concentration (g/dl)

07 Platelets ($\times 10^{9}$/l)

10 White cell count ($\times 10^{9}$/l)

AA Continuation YES NO

08 Film YES NO

11 Differential YES NO

AB Other tests YES NO

For office use only

AC Quality SATISFACTORY / SMALL CLOTS / SLIGHT GROSS LYSIS / CONTAMINATED

Delete as appropriate

Please return top copy to:
Haematology Quality Control Scheme
(DHSS-LDAG)
c/o Equine Research Station
PO Box 5, Snailwell Road
Newmarket
Suffolk CB8 7DW

Comment

Signature ____ Date ____

FIG. 2. Form used by participants for submitting results.

II. Data Processing

Processing on the computer starts when card-punching and verification is complete. It is possible to complete the statistical analyses in one working day. As the final version of each job is completed it is stored on "permanent file" until the trial report has been published. Then the entire throughput-data and results are transferred to three copies of magnetic tape for permanent storage and retrieval when required for cumulative assessments and other research studies into equipment performance, methods etc.

The objectives of this data processing are: (1) to show the overall precision of the method based on grade A weighted means and standard deviations; and (2) to provide each participant with a numerical value, known as the variance index ($|R|$) for each test or set of tests with each method.

For this several independent but linked computer programs, each with a different function, have been compiled (Fig. 3). Programs 01, 02 and 07 show the overall precision and the differences between laboratory techniques used to generate the data, and between samples of different grades. Programs 01, 02, 03 and 04 calculate the variance indices for each participant–sample–method–test combination. Program 08 prints the trial reports and programs 05 and 06 produce six-monthly cumulative assessments.

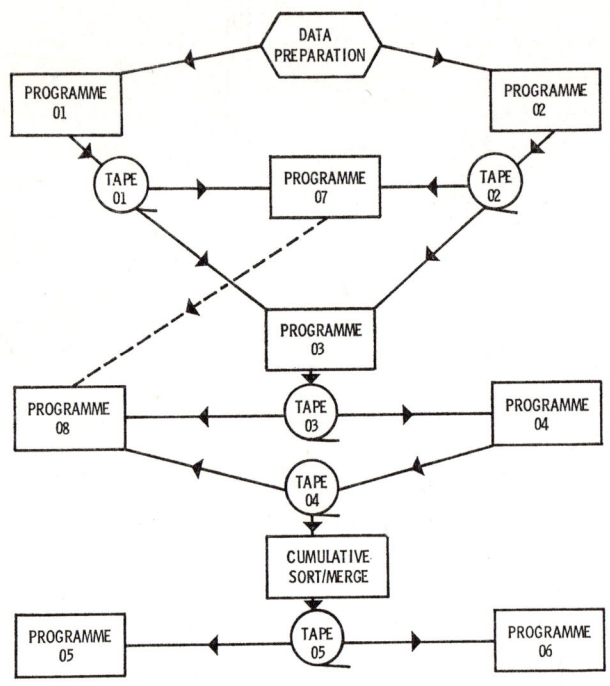

FIG. 3. Linked computer programs for processing of data in inter-laboratory trials.

Program 01—from the data, arranged as shown (Fig. 4), crude and weighted means and s.d. are calculated for every test within each sub-group. In addition $|R|$ is calculated for each test value submitted by each participant by reference to its sample-method (grade A weighted) mean and s.d.

"*Weighting*" the trial data by excluding results >2 s.d. removes any value, which is incorrect due to transcription errors or failure in equipment, methodology and/or computation. When applied to each block of sub-group test data, about one entry in every 20 is deleted. Usually there is a marked difference between the crude and weighted s.d., differences between grades suggesting that the preparations have deteriorated in the post (Table I).

"Variance index" ($|R|$) compares laboratories, methods, equipment and sample material without reference to the dimensions of the result. It is calculated for any sample–method–test group from the participant's

```
TRIAL 74-06  (JUNE,1974)  X=740601,  Y=740602,  Z=Y/X  BLOOD COUNT

  T   L   S   M     H   R   P   M     M   M   PLA  W                      SAMPLE      Q
  R   A   A   E     B   B   C   C     C   C   TE   B                      PRODUCT     U
  I   B   K   T         C   V   V     H   H   LET  C                                  A
  A       P   H                               C   S                                  L
  L   R   L   O                                                                      T
      E   E   D                                                                      T
      F                                                                              Y

  74FU01U001C  138 461 416 0600 300 320     204                          FBC0322    0A
  74FU11U001C  137 440 410 0630 310 330 156 193                          FBC0322    0A
  74FU15U001C  140 460 435 0640 300 320     162                          FBC0322    0A
  74FU35U001C  142 450 410 0690 310 340     192                          FBC0322    0A
  74FU40U001C  140 477 427 0670 288 329     185                          FBC0322    0A
  74FU74U001C  144 401 440 0978 327 325     163                          FBC0322    0A
  74FU90U001C  135 450 409 0600 306 342     164                          FBC0322    0A
  74FU93U001C  140 442 417 0650     335     136                          FBC0322    0A
  74FU98U001C  140 463 429 0620 302 326 199 197                          FBC0322    0A
               CONTINUED

  74FU53U001C  145 440 411 0600 324 352 190 191                          FBC0322    0B
  74F28SU001C  144 450 426 0650 320 338 158 198                          FBC0322    0B
  74F997U001C  141 453 459 0650 286 309 216 196                          FBC0322    0B
               CONTINUED

  74FU02U001F  135 472 390 0630 290 332 195 198                          FBC0322    0A
  74FU03U001F  141 451 412 0610 310 341     200                          FBC0322    0A
  74FU05U001F  142 447 404 0620 321 352 165 183                          FBC0322    0A
  74FU08U001F  137 398 369 0630 347 368     166                          FBC0322    0A
  74FU09U001F  138 455 412 0610 307 336 210 195                          FBC0322    0A
  74FU10U001F  142 450 413 0630 310 340     190                          FBC0322    0A
  74FU12U001F  140 457 405 0600 304 347 194 215                          FBC0322    0A
  74FU15U001F  143 448 417 0930 316 340     160                          FBC0322    0A
  74FU17U001F  138 446 412 0930 309 333     185                          FBC0322    0A
  74FU18U001F  137 460 420 0610 300 330     190                          FBC0322    0A
  74FU19U001F  135 453 410 0890 301 336     169                          FBC0322    0A
               CONTINUED

  74F016U001F  139 455 415 0900 301 333     182                          FBC0322    0B
  74FU24U001F  138 463 427 0620 298 324 214 193                          FBC0322    0B
  74FU2CU001F  140 444 423 0653 315 331 190 214                          FBC0322    0B
  74FU55U001F  138 454 426 0610 297 321 228 193                          FBC0322    0B
  74FU6BU001F  140 446 409 0610 313 337     189                          FBC0322    0B
               CONTINUED

  74FU01U002C  031 108 097 0610 285 310     063                          FBC0323    0A
  74FU11U002C  028 055 100 1020 290 280 045 057                          FBC0323    0A
  74FU15U002C  031 102 097 0620 295 340 085 061                          FBC0323    0A
  74FU35U002C  031 101 093 0620 310 330     060                          FBC0323    0A
  74FU40U002C  030 105 910 0650 282 330     063                          FBC0323    0A
  74FU74U002C  032 095 102 0689 338 321     059                          FBC0323    0A
  74FU90U002C  033 100 096 0660 328 350     054                          FBC0323    0A
               CONTINUED

  74FU53U002C  034 102 090 0650 333 378 040 059                          FBC0323    0B
  74F28SU002C  031 057 940 0650 320 330 048 063                          FBC0323    0B
               CONTINUED
  AS ABOVE TO THE END OF THE DATA.
```

FIG. 4. Raw data from trial (program 01).

test value (x) and the corresponding grade A weighted mean ($\bar{x}$) and s.d. (s_x) according to the equation:

$$|R_x| = (x - \bar{x})/s_x.$$

By definition $|R_x|$ always has a positive value and one $|R_x|$ unit is always equivalent to one s.d. (s_x) irrespective of the dimensions of the test data being assessed. Because of this lack of dimension, sample investigation ($|R_1|$) and overall ($|R_0|$) indices may be calculated for any trial covering a number of tests.

After checking for incorrect entries, the corrected data are re-run, printed out and stored on "permanent file".

Program 02—the data from program 01 are re-arranged in pairs for the same participant–method–quality group, with the card carrying the lower sample number first (x), and for each test pair within each group, the ratio ($z = y/x$) between the two samples' values (x, y) is calculated.

Table I. Inter-laboratory quality control trials (DHSS-LDAG and BCSH) trial 74-09 (August 1974) 740902 blood count

Grade		Hb (g/dl) Crude	Hb (g/dl) Weighted	RBC ×10¹²/l Crude	RBC ×10¹²/l Weighted	PCV Crude	PCV Weighted	MCV (fl) Crude	MCV (fl) Weighted	MCH (pg) Crude	MCH (pg) Weighted	MCHC (g/dl) Crude	MCHC (g/dl) Weighted	WBC ×10⁹/l Crude	WBC ×10⁹/l Weighted
A (Satisfactory 0–1 day transit)	N	105	103	104	102	103	101	103	101	104	100	102	95	104	99
	$\bar{x}$	15·54	15·65	5·19	5·23	0·462	0·466	89·01	88·84	29·95	29·93	33·60	33·62	9·26	9·44
	$S_{\bar{x}}$	0·86	0·24	0·30	0·09	0·028	0·011	1·91	1·41	0·61	0·47	0·82	0·63	0·99	0·53
	c.v.	5·53	1·55	5·87	1·78	5·968	2·433	2·14	1·59	2·05	1·58	2·43	1·88	10·70	5·60
B (Satisfactory 2–3 days transit)	N		25		26		26		26		27		28		24
	$\bar{x}$		15·62		5·17		0·465		89·98		30·27		33·42		9·32
	$S_{\bar{x}}$		0·19		0·11		0·013		1·02		0·72		0·78		0·46
	c.v.		1·21		2·06		2·820		1·14		2·38		2·34		4·94
I (lysed 0–1 days transit)	N		12		13		12		13		12		13		12
	$\bar{x}$		15·63		5·20		0·461		89·05		29·94		33·65		9·35
	$S_{\bar{x}}$		0·19		0·08		0·012		1·58		0·31		0·65		0·69
	c.v.		1·23		1·61		2·690		1·77		1·04		1·94		7·38

Table shows overall results, routine settings only, fully automated.
N = Number of laboratories, $\bar{x}$ = mean, $S_{\bar{x}}$ = standard deviation, c.v. = coefficient of variation.

This is followed by the program 01 analysis on the ratio for crude and weighted means and s.d. and the participant test ($|R_z|$) and sample variance indices. These data and results are printed out and transferred to "permanent file".

"*Ratio test*" assesses participant's values ($T = \Sigma(x, y)$) against the national means ($M = \Sigma(\bar{x}, \bar{y})$) for both samples (X, Y) for a particular test in a method group (Fig. 5) where the slope (T/M) of the line and

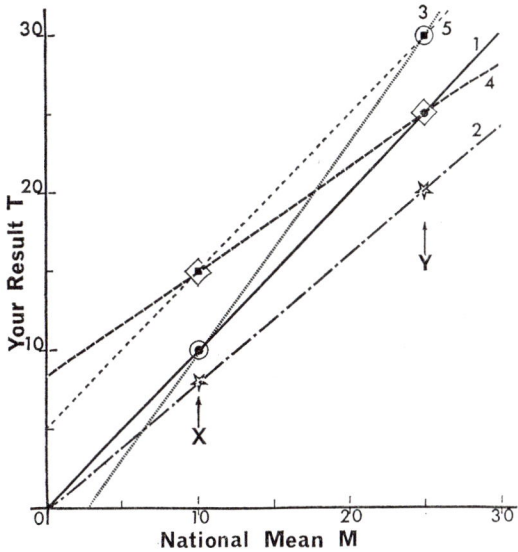

FIG. 5. Ratio test to assess participant's results by comparison with national mean.

its intercept on the T-axis show the acceptability of the participant's data. The straight lines (1–5) show some of the possible situations.

(1) For correct data from the participant (line 1)

$$y/x = \bar{y}/\bar{x} = z \text{ with } x = \bar{x} \text{ and } y = \bar{y}$$

i.e. $T = M$ and the line passes through the origin.

(2) For consistent standardization error (line 2)

$$y/x = \bar{y}/\bar{x} = z \text{ but } x \neq \bar{x} \text{ and } y \neq \bar{y}$$

i.e. $T \neq M$ and the line passes through the origin; program 01 shows both x and y to be either below or above the correct line (line 1).

(3) Displacement of the T-axis intercept from zero (line 3) and wrong alignment (lines 4 and 5) caused by X or Y both being in error (program 01) results in $z = y/x$ being either greater or less than its true value ($\bar{z} = \bar{y}/\bar{x}$) (program 02).

Program 03—the results from program 01 and 02 are sorted into increasing participant laboratory number and simultaneously the unsatisfactory data are eliminated from the permanent file in readiness for the calculation of the investigation ($|R_i|$) and overall ($|R_o|$) indices by program 04.

```
TRIAL 74-04 (APRIL 1974) X = 740401, Y = 740402, Z = Y/X. (FULL BLOOD COUNT)
HISTOGRAM FOR VARIANCE INDEX METHOD F
(/R/)  0.0   0.2   0.4   0.6   0.8   1.0   1.2   1.4   1.6   1.8   2.0   24.07 MAXIMUM
LAB NOS
         0    50     5     2    16    15     4     3    30     0          26
         0   183     9    20    17    24    75    31   133     0         181
         0   192    18    34    19    29    90    64   204     0         214
         0   226    46    36    22    32   100    68     0     0         220
         0     0    48    40    25    56   159    95     0     0         294
         0     0    61    41    28    57   200   103     0     0         259
         0     0   129    47    43    62   230   128     0     0         314
         0     0   130    54    45    67   241   191     0     0         319
         0     0   131    58    49    72   242   236     0     0         321
         0     0   137    59    63    84     0   246     0     0         335
         0     0   138    65    78   102     0     0     0     0           0
         0     0   150    77    85   126     0     0     0     0           0
         0     0   190    82    97   158     0     0     0     0           0
         0     0   195    86   107   213     0     0     0     0           0
         0     0   199    88   108   247     0     0     0     0           0
         0     0   219    89   114   298     0     0     0     0           0
         0     0   225    92   116   299     0     0     0     0           0
         0     0   250    96   117     0     0     0     0     0           0
         0     0   262   109   119     0     0     0     0     0           0
         0     0   263   111   135     0     0     0     0     0           0
         0     0   276   112   143     0     0     0     0     0           0
         0     0   333   121   193     0     0     0     0     0           0
         0     0     0   125   198     0     0     0     0     0           0
         0     0     0   134   207     0     0     0     0     0           0
         0     0     0   139   222     0     0     0     0     0           0
         0     0     0   145   243     0     0     0     0     0           0
         0     0     0   149   256     0     0     0     0     0           0
         0     0     0   151   268     0     0     0     0     0           0
         0     0     0   164   275     0     0     0     0     0           0
         0     0     0   165   296     0     0     0     0     0           0
         0     0     0   167   999     0     0     0     0     0           0
         0     0     0   182     0     0     0     0     0     0           0
         0     0     0   186     0     0     0     0     0     0           0
         0     0     0   197     0     0     0     0     0     0           0
         0     0     0   205     0     0     0     0     0     0           0
         0     0     0   253     0     0     0     0     0     0           0
         0     0     0   266     0     0     0     0     0     0           0
         0     0     0   305     0     0     0     0     0     0           0
         0     0     0   306     0     0     0     0     0     0           0
         0     0     0   323     0     0     0     0     0     0           0

COUNT      0     4    22    40    31    17     9    10     3     0    10 TOTAL = 146
% TOTAL 0.00  2.74 15.07 27.40 21.23 11.64  6.16  6.85  2.05  0.00  6.85

                                                   CRUDE      WEIGHTED

VARIANCE INDEX (/R/) MEAN                       1.19 UNITS    .99 UNITS

                     STANDARD DEVIATION         2.05 UNITS    .65 UNITS

                     COEFFICIENT OF VARIATION % 173.17        65.52

                     NO IN GROUP                146           144
```

Fig. 6. Histogram of variance index for blood count by fully-automated equipment.

Program 04—only satisfactory data, i.e. grades A, B, C and D are used to calculate ($|R_i|$) and $|R_o|$ by the formulae:

$$|R_i| = (|R_x| + |R_y| + |R_z|)/3$$

$$|R_o| = \left\{ \sum_{n=1}^{n=k} (|R_n|) \right\}/n.$$

$|R_n|$ is the $|R_x|$, $|R_y|$ or $|R_z|$ of each and every test and k the total number of indices (usually three for each test and 21 for a method when platelets are omitted) for each method returned by each participant. If only one sample was satisfactory $|R_i| = |R_x|$ and $|R_o|$ is the sample index. $|R_o|$ assesses the participant's performance in a trial for a method and each method is assessed independently.

This program also prints the national distribution histogram of $|R_o|$ for each method. In these histograms each laboratory is identified by its reference number (e.g. Fig. 6) so that a rapid check can be made to assess the situation when a laboratory returns an exceptionally good performance or one which is unsatisfactory. Usually the former ($|R_o| < 0.4$) is due to a participant's returns showing many tests without data; the latter ($|R_o| > 2.0$) to transcription errors. For both $|R_i|$ and $|R_o|$ data to be available for equipment and technique studies when required and for the trial report (program 08) they are stored on "permanent file".

Program 07 calculates the overall grade A sample and ratio means, s.d. and standard errors of the mean (s.e.) using the "weighted" data. These results are sent to participants who do not return data in the trial before closing date, to enable them to check their results.

Distribution analyses of the sample and ratio data of all satisfactory specimens are also prepared by this program. These analyses (Fig. 7)

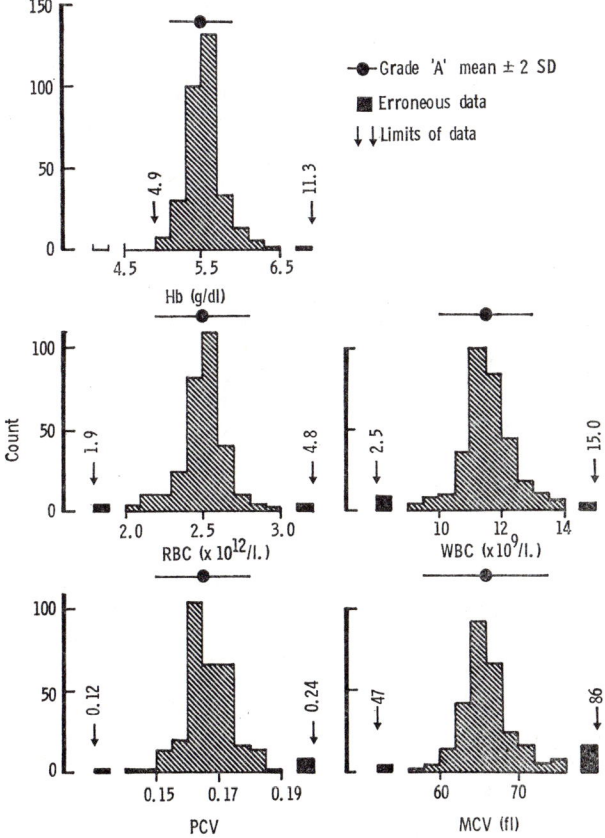

Fig. 7. Distribution analysis of blood-count results for all satisfactory specimens, compared with grade A data; sample 740301.

are constructed for Hb, RBC, WBC and platelet counts, as these are directly subject to instrument and method variation and not to computation error, and also for PCV and MCV which are interdependent with one experimentally derived measurement. The overall grade A mean is the mid-point of each plot; 2 s.d. covers about five increments of distribution on either side of it. The outermost and sixth distribution on either side covers the rest of the range of the data from the minimum to maximum values reported.

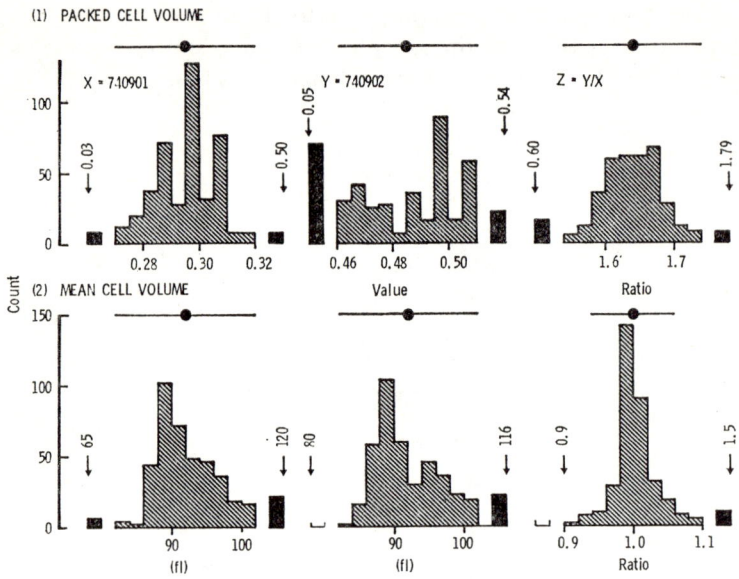

Fig. 8. Distribution of PCV and MCV results on two samples (x, y) and their ratio (z). Variation in preparations are shown for trial 7409.

The modal variations in these plots reflect the suitability of the preparations for all method assessments. A sharp mode shows the product was suitable for any equipment and any method whilst a bi- or multi-modal distribution stresses that the preparation has limited use. The PCV and MCV distributions provide useful information on this (Fig. 8).

Program 08 produces the trial report from the data now stored of "permanent file". A report (Fig. 9) is printed for each consecutive laboratory number and contains: (1) the national means and s.d. of the methods used by the participant, extracted from program 01 and 02 results or from program 07 if data was not submitted; (2) the participant's data and indices, from program 01, 02 and 04; (3) comments on the trial, and (4) plans of future trials.

At present, acceptability of performance is reported in two ways so a participant may interpret his data in terms of the national mean ($\bar{x}$) and s.d. (s_x) for a given test by a particular method, and also in terms of $|R|$. The trend in performance from one trial to the next may

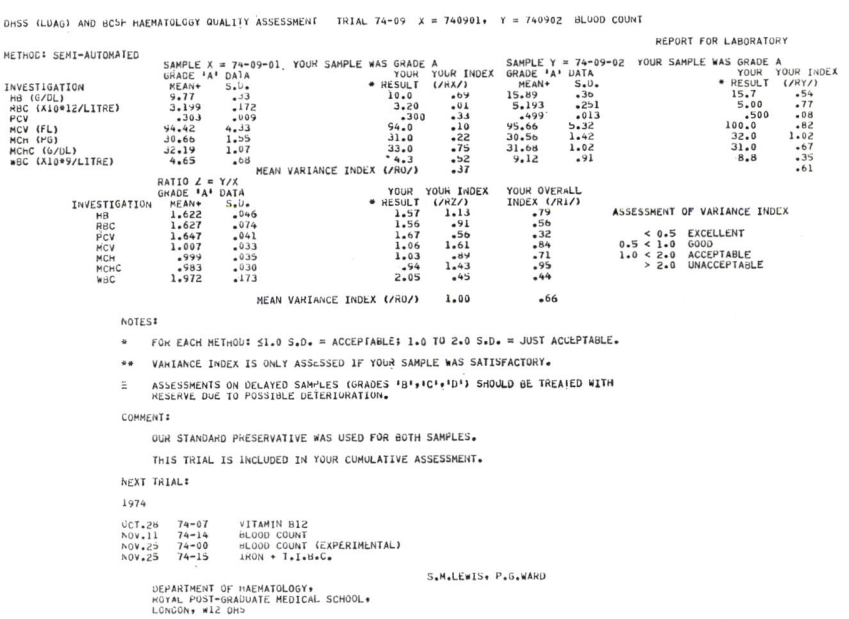

FIG. 9. Print-out of trial report as issued to a participating laboratory.

be followed with $|R|$ and an appreciation of its relationship to changes in the sizes of $\bar{x}$ and s_x keeps all data in perspective (Table II).

Finally, when all the reports have been printed, the information on "permanent file" is transferred to "magnetic tapes".

III. CUMULATIVE ASSESSMENT

All the data, stored on magnetic tapes, are used in this assessment. Processing by two programs—05 and 06—assesses the changes of performance at each laboratory and provides a picture of the national trend, between two consecutive trials and between two chronological periods.

Program 05 calculates the mean ($|R_c|$) and s.d. ($|R_s|$) of the overall ($|R_o|$) indices for each method for each participant in each period

Table II. Variance index changes between successive trials

Index ($|R_n|$)*	$Dx_n = (x - \bar{x})_n$	$(s_x)_n$	Effect
No change $\quad$ $|R_2| = |R_1|$	No change	No change	No change
Decrease $\quad$ $|R_2| < |R_1|$	(a) No change	$(s_x)_2 > (s_x)_1$	Possible improvement
	(b) $Dx_2 < Dx_1$	No change	Marked improvement
Increase $\quad$ $|R_2| > |R_1|$	(a) No change	$(s_x)_2 < (s_x)_1$	No improvement
	(b) $Dx_2 > Dx_1$	No change	Deterioration
$|R|$ = minimum	No change	s_x/x = minimum	No improvement
$|R| = 0.0$	(a) $x = \bar{x}$	No change	No improvement
	(b) $x = \bar{x}$	s_x/x = minimum	No improvement possible
	(c) $x = \bar{x}$	$s_x = 0$	No improvement possible

* For first trial $n = 1$ and for the next trial $n = 2$.

irrespective of the amount of data received. It prints distribution histograms of $|R|$ for each method provided that the minimum number of trials have been reported for both periods.

Program 06 compares performances. In this each participant receive a cumulative assessment of performance throughout the period. This provides: (1) a plot of $|R_o|$ against trial for each method (F, M and S); (2) the weighted $|R_c|$ and $|R_s|$ for each period and method, with the number of trials where $|R_o| > 2.0$ stated, and (3) statistically derived comments on any change of performance between periods A and B and on performance within period B.

Biennial reviews are undertaken of the accumulated data. These show that performance assessment of individual laboratories requires the following factors to be taken into account: (1) insufficient data received due to intermittent participation, (2) late entry into the scheme by the participant, (3) introduction of new methods by a participating laboratory, (4) occasional abnormally high values of $|R_o|$ due to transcription errors, and (5) marked changes in laboratory performance.

Maximum information is obtained by dividing the data into two chronological periods (A and B) with an equal number of trials in each group. This will help to identify factors associated with (1), (2) and (3). Weighting the $|R_o|$ data, to exclude a result if $|R_o| >$ (crude $|R_c|$) + (crude $|R_s|$), will identify and overcome the problem of transcription error.

Good performance by a single laboratory in all methods is rare, i.e. $|R_t| < 1.0$, $|R_c| < 1.0$ and $|R_s| < 0.5$ (Fig. 10) where $|R_t| = |R_c| + |R_s|$. Good performance in two of the methods is more common and average

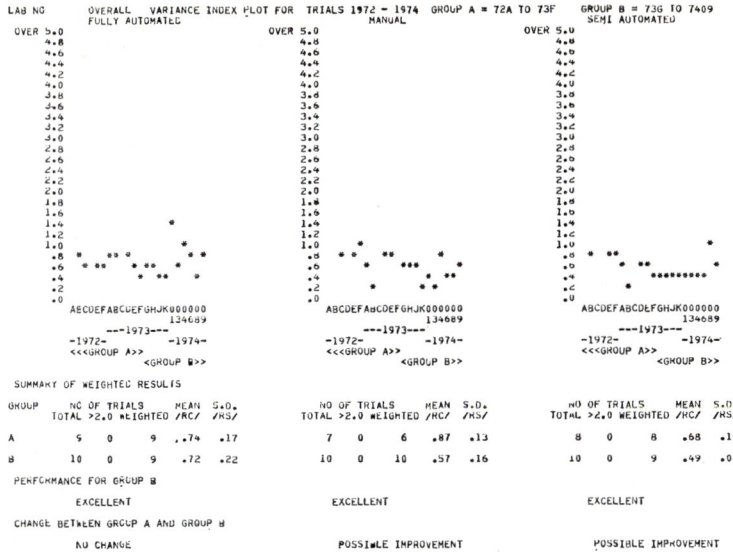

Fig. 10. Cumulative assessment of performance; see text.

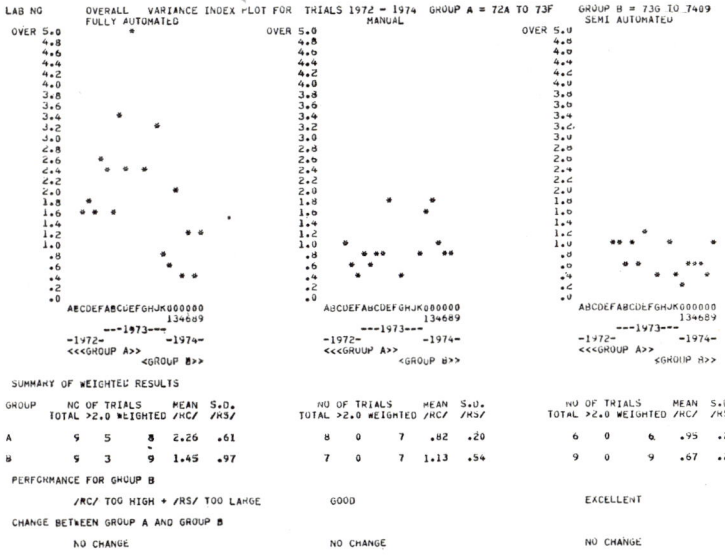

Fig. 11. Cumulative assessment of performance; see text.

performance in two of them is usual. A few laboratories continue to be consistently poor, i.e. $|R_t| > 2.0$, $|R_c| > 2.0$ and/or $|R_s| > 1.0$ (Fig. 11). The national pattern is summarized in Table III, for changes both between periods and within period B.

TABLE III. Trends in laboratory performance from trial 72A to 7409 as shown by acceptability index $|R_t|)$

Acceptability index ($|R_t|$) Period	Fully-automated A* No.	B† No.	Manual A No.	B No.	Semi-automated A No.	B No.
Excellent						
0·0–0·5	0	0	0	0	0	0
0·5–1·0	31	53	33	3	43	54
Good						
1·0–1·5	54	65	50	15	76	53
1·5–2·0	12	11	20	14	24	23
Poor‡						
2·0–2·5	4	6	9	2	13	11
2·5–3·0	5	4	5	1	6	1
3·0–3·5	2	1	—	—	2	1
3·5–4·0	1	1	—	—	—	1
4·0–4·5	—	—	—	—	—	—
4·5–5·0	—	—	—	—	—	1
>5·0	—	4	—	—	—	—
Total	109	145	117	37	164	145

* Group A covers trials 72A to 73F (12 trials).
† Group B covers trials 73F to 7409 (10 trials).
‡ The laboratories in these categories are told if their $|R_c|$ (mean $|R_0|$) is too high and/or their $|R_s|$ (scatter about $|R_c|$ of $|R_0|$) is too large.

IV. CONCLUSIONS

This scheme has become well established, and participants have become familiar with the print-outs.

Frequent program revision has been necessary, particularly for program 08, which prints the trial reports, so that the information sent to participants can be made as clear and concise as possible.

Furthermore, trials have revealed pecularities in blood preservatives, equipment and methods which have had to be taken into account during processing to achieve fair assessment of data. Revision of programs 05 and 06 has also been necessary before each cumulative assessment, because of the increasing number of trials which have been distributed.

A library of computer programs has been compiled to process trial data and assess laboratory performance. These might be of use to other schemes of a similar nature.

REFERENCE

Lewis, S. M. and Burgess, B. J. (1969). *Br. med. J.* **4,** 253–256.

4. Inter-Laboratory Trials: The Quality Control Survey Programme of the College of American Pathologists

JOHN A. KOEPKE

Department of Pathology, University of Iowa College of Medicine,
Iowa City, Iowa, U.S.A.

I. THE COLLEGE OF AMERICAN PATHOLOGISTS SURVEY PROGRAMME

The College of American Pathologists (CAP) has been active in survey programmes in laboratory work for more than a decade. This programme initially began with proficiency testing in chemistry but was rapidly expanded to include microbiology and haematology as well as a larger series of so-called special surveys. The first haematology survey was conducted in 1962 and measured the inter-laboratory variability in haemoglobinometry (Brown *et al.*, 1962).

At the present time the programme in haematology has evolved to include quarterly kits each containing about 100 tests. These include specimens for haemoglobin (Hb), haematocrit (PCV), red-cell (RBC) and leucocyte (WBC) determinations, a coagulation panel including prothrombin time, partial thromboplastin time and fibrinogen determinations and a series of transparencies and unstained blood smears for evaluation of proficiency in morphologic haematology. The kits also contain a urinalysis specimen and a single transparency for urine microscopic examination. The kits are manufactured by Hyland Laboratories, Costa Mesa, California under the direction of the Survey Committee of the CAP. About 2400 laboratories are enrolled in the comprehensive haematology programme. A larger group of laboratories, about 4400, receive a smaller set of haematology specimens for analysis, the so-called basic survey (CAP Survey Manual, 1974). Thus most American hospital laboratories participate in these surveys.

A variety of different haematological skills are tested in the programme. Each skill calls for different methods of specimen preparation, data analysis and evaluation. These factors will be discussed in the paragraphs which follow.

TABLE I. Haematological skills tested in the programme

Skill	Test
Quantitative haematology	
Particle counting	RBC, WBC
Chemical methods	Hb, fibrinogen*
Physical methods	PCV
Coagulation procedures	Prothrombin time, partial thromboplastin time,* fibrinogen*
Morphology	Photomicrographs,* blood smears

* Not included in the basic survey.

The objectives of the programme in haematology include the determination of the "state of the art" (Koepke, 1974). The surveys also reveal controversial areas. After these problem areas have been identified, we are interested in promoting standardization of procedures and in encouraging more uniformity of nomenclature for reporting cell types. Finally the programme has become involved in the measurement of proficiency and acceptability of the laboratory's performance for the inspection and accreditation programmes of the CAP as well as governmental agencies such as the Center for Disease Control (CDC) or individual state departments of health.

II. QUANTITATIVE HAEMATOLOGY

A. THE HAEMOGLOBIN MODEL

The programmes in quantitative haematology have indicated a continuing improvement in the measurement of all red-cell parameters.

This is best exemplified by the evident improvement over the years in the performance of Hb measurements. The provision and use of certified standards have had a great deal to do with this improvement (Eilers, 1965).

Figure 1 illustrates haemoglobin precision over this time period. In general the coefficient of variation (c.v.) of 4–5% found in 1962

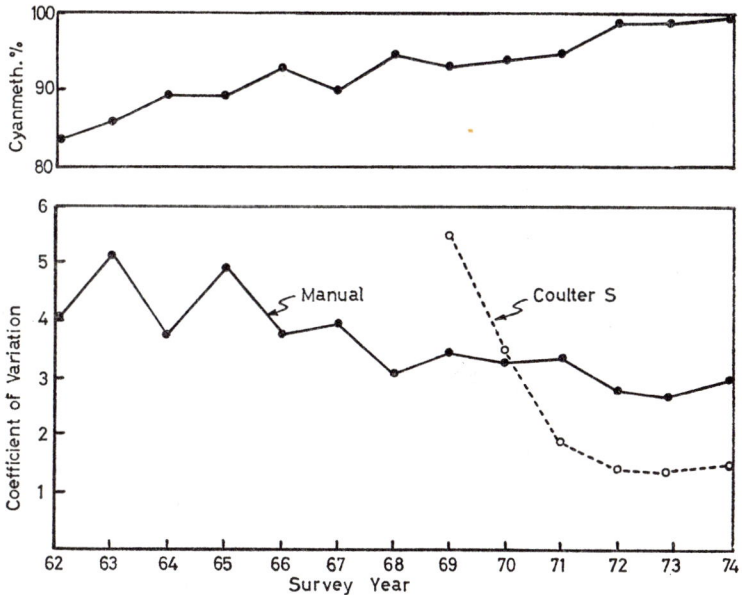

FIG. 1. Haemoglobin, precision, 1962–74.

dropped to about 2·5% in 1973 when using manual techniques. In the recent years there has been a great increase in automation in haematology laboratories. The surveys have accumulated a significant amount of data on the Coulter S instrument. These show an even more striking improvement in haemoglobinometry, especially in the last 4 years (Fig. 1). It cannot be said with certainty whether we have come to an irreducible minimum in variability for both the manual and automated techniques but there appears to have been a levelling off over recent years.

Interestingly, during the 13-year period there has been a gradual coincidental increase in the number of participants using the cyanmethaemoglobin procedure (Koepke, 1973). This indicates that method standardization and the availability of certified standards due to this levelling off of precision has made a very important contribution to haemoglobinometry precision.

The Survey Committee has more recently initiated absolute levels of precision for haemoglobinometry evaluations. We have used ± 0.4 g/dl as the limits of good performance and ± 0.8 g/dl as the limits of acceptable performance. When one determines what has been the actual performance by both the comprehensive and basic laboratory participants one is able to see that these evaluation limits are both reasonable and readily attainable at the present time (Fig. 2). It can

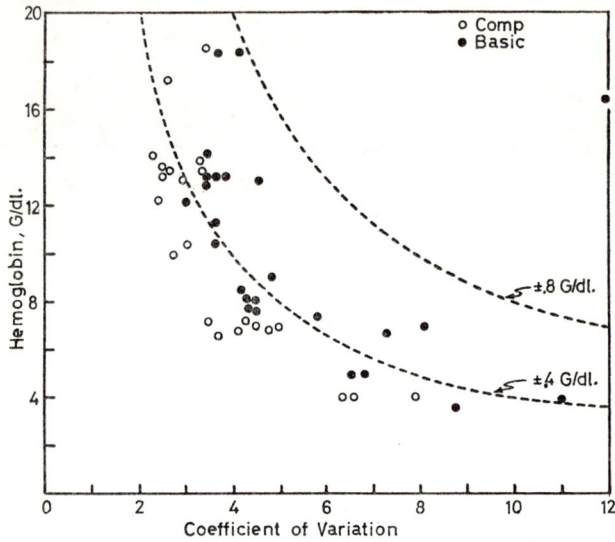

FIG. 2. Haemoglobin, precision, 1970–72.

be seen that the mean levels of all except one Hb unknown in 1971 are within these limits. The one specimen found to be unduly variable (A-16-1972) was found to give a significant problem in aliquoting during the manufacturing process. Essentially then, the participant performance functions as a quality control programme for survey specimens, when participant results indicate a survey specimen is more variable than expected, these results are not suitable for laboratory performance evaluation.

These studies on the haemoglobin model indicate that we should now direct our efforts towards improving laboratory performance in other procedures. Probably no significant increase in medical usefulness can be gained by further improvement in precision in haemoglobinometry.

B. PARTICLE COUNTING

In 1965 a large group of laboratories (1626) participated in the CAP comprehensive hematology survey. The results of this survey were presented at a session of the European Society of Haematology devoted to standardization in haematology (Koepke and Eilers, 1966). After several years, the complete red-cell parameter survey programme was reinstated and has continued to the present. Representative data comparing the performance of the several methods for red-cell particle counting are tabulated in Table II.

TABLE II. Red-cell counting precision*

Method	1965	1969	1970	Survey year 1971	1972	1973	1974
Haemocytometer	10·8	10·2	8·4	8·2	8·5	8·0	7·4
Coulter (single channel)	9·0	6·2	5·7	5·6	6·2	6·3	5·3
Coulter Model S	—	3·0	2·9	2·0	2·4	2·1	2·0
Fisher Hemalyzer	—	8·1	7·3	8·7	10·1	7·0	3·5
Technicon SMA 4/7	—	6·3	5·2	7·7	6·3	6·7	5·1

* Expressed as coefficients of variation of normal range red-cell counts.

In quantitative procedures, the survey committee analyses the data in the following manner: all data are examined and a mean and standard deviation are calculated. All values lying outside of ±3 s.d. are classified as "outliers" and excluded from subsequent analyses. The remaining data is re-analysed and an adjusted mean and s.d. are determined. In evaluation of a laboratory's performance, all values between ±1 s.d. of the mean are considered "good" performance; those between ±1 and ±2 s.d. are called "acceptable"; values outside of ±2 s.d. are "not acceptable".

C. HAEMATOCRIT (PCV)

The performance in the manual microhaematocrit procedure shows an evident improvement since 1965 when the coefficient of variation (c.v.) was 5·5% on a specimen with an elevated PCV (0·62). By 1971 a 3·1% c.v. was noted and by 1974 a 2·5% c.v. was found.

III. COAGULATION PROCEDURES

A. PROTHROMBIN TIME

The survey programme in coagulation has also documented improvement in performance of the prothrombin determinations over the years.

In 1963 the College of American Pathologists conducted a rather extensive and complete survey of the prothrombin time test in more than 1000 laboratories. These studies investigated many aspects of this test including methods, type of thromboplastin, and other variables. The report of the 1963 survey found, for example, that the ten available commercial thromboplastins varied in reactivity, making comparisons between different thromboplastins very difficult due to the wide variations obtained in the prothrombin times. Several of these thromboplastin reagents were, in fact, felt to be not acceptable due to the unusually wide variability when blind replicate specimens were tested. (Miale and LaFond, 1967).

The CAP survey of prothrombin time proficiency carried out in 1969 documented a number of changes in the performance of this test (Koepke, 1970). Firstly there was a significant improvement in overall precision of prothrombin time testing. Also, most of the thromboplastins shown to be poor in the 1963 survey were no longer marketed; others were improved.

Semi-automation of this test made significant inroads in the clinical laboratory. Whereas no mechanical devices were used in 1963, about 40% of laboratories used such an instrument in 1969. The Fibrometer (BBL) was by far the most popular instrument. At present instruments are used in 74% of all reporting laboratories. This factor is most likely responsible for the improvement noted, since comparable results of the Fibrometer with the tilt tube or loop procedure shows evident greater precision using the automated method.

A number of conclusions were drawn from the examination of the 1969 prothrombin survey data.

(1) The participants in the basic and comprehensive surveys appear to do prothrombin time determinations equally well.

(2) No consistent pattern of thromboplastin reproducibility emerged, indicating that all major American thromboplastins were essentially equal in quality.

(3) There are, however, consistent biases in the prothrombin time as measured by the several commercial thromboplastins. The greatest differences are more than 4 s in the "therapeutic range" primarily being related to the raw material used for the preparation, i.e. brain or brain–lung preparations.

Since 1969, there has been a levelling off in the prothrombin time precision (Table III). Some recent increased variability of prothrombin testing procedures has been due to the introduction of several newer thromboplastins into the marketplace. The mean times of the newer products have been different by about 1 s in the normal range. Until 1974 the survey questionnaire did not separate these different products

TABLE III. Prothrombin time precision*

Survey year	Normal range (12–14 s)		Elevated range (24–28 s)	
	Manual	Automated	Manual	Automated
1963	10·7	—	15·3	—
1969	11·0	7·3	8·0	7·0
1970	6·7	5·9	7·1	7·7
1971	7·6	5·7	8·9	8·2
1972	7·5	5·5	8·6	7·4
1973	8·9	6·4†	8·6	7·5
1974	7·8	6·7†		

* Expressed as coefficients of variation.
† See text for explanation.

from the same manufacturer and hence a higher c.v. was calculated. The current survey statistics indicate that indeed the newer product is quite acceptable with a c.v. of around 4·6% using automated methods. This approaches the best precisions seen in our survey programme and indicates that improvement in the performance of the prothrombin time test is continuing.

B. PARTIAL THROMBOPLASTIN TIME

The partial thromboplastin time (PTT) has become a keystone procedure in clinical laboratories throughout the country. A PTT specimen was first included in the 1969 CAP haematology survey, and since 1970, four PTT specimens have been included each year.

In 1969 it was apparent that the measurement of the partial thromboplastin time left much to be desired. Coefficients of variation ranged from 20–57% for various PTT systems. Since 1969, significant changes have been observed in this determination. For example, there has been a significant trend towards activation of the PTT test. In 1969, 69% of the laboratories used identifiable activated procedures, whereas in 1973, 95% of the laboratories were using an activated method.

A study of survey returns indicates that four major methods are used with five major manufacturers of PTT reagents. From this matrix, at least 20 different systems are possible, but in fact, ten PTT systems account for 70% of all reporting laboratories, and three systems (Dade/Fibrometer, General Diagnostics/Fibrometer, and Dade/tilt tube) account for just over half of the PTT systems in use this year. As has occurred in the measurement of the prothrombin time, there has been a marked trend towards automation of the PTT in clinical laboratories.

Indeed, two thirds of the laboratories are using the Fibrometer at this time. Another 17% use the tilt tube methods. A variety of other instruments are also being infrequently used for this measurement.

From the CAP survey data, we can compare the precision of all of these methods in various combinations with all the PTT reagents (PTT systems). The techniques divide into two major groups; systems using activated procedures which have c.v. between 13 and 22% and the non-activated procedures which have significantly higher c.v. The increased precision of the activated procedures is not easily explained since four different methods and four different partial thromboplastin reagents are in common use. The three most precise methods are "total systems" in which the manufacturer of the instrument and of the reagent is, in fact, the same organization, or one in which a close working relationship was established. This co-operation has resulted in better precision of this particular measurement (Table IV).

TABLE IV. Systems for partial thromboplastin time, ranking by precision

Partial thromboplastin	Method	(%)
Hyland	Clotek	13·3*
BBL	Fibrometer	15·4
Dade	Electra	16·6
General Diagnostics	Fibrometer	16·8
Hyland	Fibrometer	17·4
General Diagnostics	Tilt tube	17·8
Dade	Tilt tube	19·7
Dade	Fibrometer	22·3
Ortho (non-activated)	Tilt tube	32·2
Ortho (non-activated)	Fibrometer	49·9

* Figures indicate coefficient of variation of system.

The accuracy of the PTT is a much more difficult question. A number of tentative conclusions can be drawn, however. The survey committee evaluates the performance on the partial thromboplastin time in a very straight-forward manner. The following system has been used. The participant reports the PTT on the unknown survey sample as well as the upper limit of normal used in their laboratory for the partial thromboplastin time. For each group of laboratories using a similar system (manual/activated, manual/non-activated, automated/activated, and automated/non-activated) the evaluation consists of comparing the participants results with the upper limit of normal for his laboratory. The result then is normal or abnormal using these limits. Results (i.e. normal or abnormal) submitted by 80% or more of the participant laboratories using the same type of reagent and technique define acceptable performance for that method. When less than 80% agree, the results are not evaluated.

Of the systems used for reporting results, the manual methods are generally more sensitive to mild abnormalities of the PTT than the most popular mechanical procedures presently in general use. Since 1971, each of the 12 PTT unknowns have been analysed for levels of AHF in the lyophilized survey specimens. AHF levels ranged from 214%–0·6% AHF. The ten most popular systems yielded surprisingly good linearity over the entire range of AHF concentration.

Several recommendations seem warranted after study of the survey data. (1) Because of the poorer precision of non-activated procedures, they should be eliminated. (2) Since there is an apparently improved performance in several PTT *systems*, such a system may prove to be a good choice for an individual laboratory. (3) Because of the inherent variability of the systems, each laboratory should determine its own normal PTT range, particularly the upper limit of normal. The upper limit of normal is defined as a 50% AHF level without any other coagulation factor deficiencies (Koepke, 1975).

C. FIBRINOGEN

The first assessment of fibrinogen measurements was made in 1967 when a single lyophilized citrated plasma specimen was included. Results indicated that a significant problem existed with wide c.v. for all methods (Table V). In addition, a great variety of methods were

TABLE V. Fibrinogen measurements

Method	1967	1971	Survey year 1972	1973	1974
Parfentjev	66*(33)†	42(34)	30(31)	21(27)	16(33)
Biuret	6(27)	3(37)	3(28)	2(24)	1(28)
Phosphate buffer	—	3(34)	2(24)	2(21)	2(42)
Ellis–Stransky	—	1(31)	2(24)	2(22)	1(26)
Thrombin time (Dade)	—	—	50(11)	62(11)	70(11)
Other quant.	21(31)	41(42)	9(34)	14(41)	5(68)
Semi-quant.	7(28)	10(130)	4(90)	3(455)	7(84)

* Per cent of laboratories using the method.

† Figures in parentheses indicate method c.v. of fibrinogen specimens in normal range.

being used, indicating the lack of a really satisfactory technique for this measurement.

In 1971 four fibrinogen specimens were again included and this pattern of four specimens per year has continued to the present. No significant improvement was evident with any of the methods. However, during this year (1971) a commercial (Dade) modified thrombin time for fibrinogen measurement was placed on the market. A startling improvement in fibrinogen precision was evident and this fact, coupled with the simplicity of the procedure using semi-automated techniques (usually the Fibrometer) has apparently resulted in a rapid increase in the use of this new method. Almost three-quarters of the laboratories have now switched to this method and the change seems to be well founded. The older methods continue to show the variability first observed in 1967 (Table V).

In 1974, in a co-operative venture with the haematology group at the Center for Disease Control (CDC), a pair of specially prepared fibrinogen specimens were sent to participants. A measured increment of fibrinogen was added to half of a parent plasma pool. The two pools were aliquoted and lyophilized. Preliminary evaluation of this data indicates that the Dade procedure, in addition to its proven precision, also seems to most accurately measure the incremental fibrinogen levels (Koepke *et al.*, 1975).

IV. Morphological Haematology*

The CAP surveys have been working to improve communications between morphologists for about 10 years. This portion of the programme causes greater frustration and more controversy (as well as generating extremely worthwhile data) than most other areas in the programme. Morphological identification touches on the professional competence and personal expertise of the survey participants, and therefore is predictably controversial. The objectives of the morphological haematology surveys are similar to the other sections of the programme as previously noted. But first of all, we have been interested in determining the "state of the art" in morphological haematology.

The informal nature of the earlier CAP survey programmes has been changed by the advent of accreditation programmes for laboratories in the United States. These regulations require that a laboratory perform acceptably in a proficiency testing programme. Competence in a field such as morphology is considerably more difficult to assess than in quantitative procedures. For example, consider the difficulties

* This topic is also discussed in chapter 11.

one might experience in determining competence in cytology or surgical pathology. With the possibility of revoking a laboratory's license, significantly increased numbers of complaints and associated controversy ensue regarding this part of the programme. This was anticipated and has for the most part been quite productive.

In any surveys, some ground rules must be set for the participants. With this in mind the CAP Survey Manual includes a hematology glossary in which normal peripheral blood and nucleated bone marrow cells are described in some detail. Survey participants are asked to use these definitions in their identification of haematological cells in the survey specimens. The hematology glossary, authored primarily by Dr L. W. Diggs, represents a reasonable concensus for normal haematology cells (CAP Survey Manual, 1974).

The programme in morphological hacmatology has been divided into two major types of study. The first is the 35 mm photomicrograph sent only to comprehensive hematology survey participants and the second portion is the blood smear evaluations sent to all participants. The transparencies have primarily been the work of Dr Diggs who photographed many of them. These slides are chosen for their suitability in survey work. They are circulated to a reviewing board of ten haematologists known for their expertise in morphological haematology and the identification concensus of this group constitutes "good" performance by the survey participants. In addition, however, review of the survey data may provide for additional "acceptables" if this is felt to be necessary.

Up to the end of 1973 a total of 115 different transparencies were sent to survey participants (Table VI). They were first included in the 1965 comprehensive hematology survey and have been included each year since that time (Koepke, 1966). As can be seen from the table

TABLE VI. CAP survey transparencies, 1965–73

| | Cell type | | | |
Abnormality/specimen	Leucocyte	Erythoclyte	Megakaryocyte, other	Total
None, peripheral blood	20	0	0	20
None, bone marrow	22	11	17	50
Common peripheral blood abnormalities	9	12	2	23
Uncommon, rare, bone marrow	16	2	4	22
Total	67	25	23	115

approximately 20 pictures were of normal peripheral blood constituents; 50 were normal bone marrow cells including leucocytes, red-cells and other marrow elements. Twenty-three cases of common red-cell and leucocyte abnormalities seen in peripheral blood constituted a third grouping and a small group of 20 uncommon or rare haematological abnormalities completes the survey photomicrograph set up to the end of 1973.

As would be expected, the normal leucocytes in the peripheral blood are identified quite well with a very good percentage correct for these common cells (Table VII). Some significant differences of opinion

TABLE VII. Normal leucocyte identification, peripheral blood

Leucocyte	Correct identifications (%)
Segmented neutrophil	99
Eosinophil	96
Basophil	95
Monocyte	87
Lymphocyte	96

regarding the differentiation of segmented from stab neutrophils have become apparent. A clinical-laboratory study delineating the clinical usefulness of the rather arbitrary identification criteria indicates the rather rigid definitions for segmented and stab neutrophils are quite helpful in discovering early or mild inflammatory conditions in patients with normal total white counts (Mathy and Koepke, 1974).

It is expected that common red-cell abnormalities seen in peripheral blood would be correctly answered by the survey participants. Table VIII shows an almost unanimous correct identification rate of sickle

TABLE VIII. Common red-cell abnormality identification, peripheral blood

Red-cell abnormality	Correct identifications (%)
Sickle cells	99
Target cells	99
Rouleaux of red-cells	99
Microcytic, hypochromic erythrocytes	97
Howell–Jolly bodies	92

cells, target cells and rouleaux. However, predictably, the identification of uncommon cells or cells ordinarily seen only in bone-marrow preparations leads to more difficulties for the survey participants. Also controversial are the unusual or rare phenomenon such as metastatic carcinoma (Keitges and Koepke, 1971).

The advantage of the transparencies is that all participants are looking at the same cell which avoids problems such as field selection and variable staining. There are also some disadvantages. Many participants feel "confined" by the inability to scan the slide; many also wish to stain the slide themselves. The earlier surveys used only blood smears in morphological haematology which were later abandoned. Peripheral blood smears were reinstituted in 1970 and continue to be an integral part of the hematology survey.

Several years ago a questionnaire was sent out with a stained blood smear. We asked these people to compare the stain on the survey slide with the one to which they were accustomed. More interestingly, when we compared results of our staining of red-cells and leucocytes, those who preferred receiving stained smears generally said that our stain was much the same. Those who preferred the unstained smears felt the survey stained slide was significantly *bluer* or *redder* than their own smears. Likewise, the leukocytes apparently were more darkly stained than most people were accustomed to, but this is probably inherent with the use of the Ames slide stainer. There was no clear concensus as to preference on how slides should be stained. This undoubtedly leads to some of the problems that are generated in the blood smears.

The preparation of more than 30 000 blood smears annually is indeed a significant logistic problem. While appropriate and interesting patients can easily donate a tube of blood to make up several hundred smears, the discovery of an appropriate non-anaemic patient who is able and willing to donate the 150 ml of blood needed for each basic survey is difficult. To make 4500 satisfactory slides within several hours after drawing the specimen and just a few weeks before the smears are shipped to participants becomes difficult on occasion.

Results on the differential leucocyte counts used in the surveys have been most interesting. For example, it is of interest to see that the survey participants closely match predicted sampling variabilities insofar as identification of common cell types seen in peripheral blood. The 95% ranges predicted by Rümke (1960) have proven to be quite accurate in the surveys. The correlations are amazingly good for these common leucocytes found in peripheral blood. Atypical lymphocytes and other unusual cells are less well identified in the same preparations.

We have on occasion sent smears stained for reticulocytes. A number of years ago a patient with a profound haemolytic anaemia and a very elevated reticulocyte provided the source for survey blood. An unduly wide variability in the reticulocyte percentages (40 ±20%) was reported. It was apparent that many laboratories do not identify a reticulocyte if it has just one granule of reticulin in the cell. In this particular case, almost 30% of the cells in the smears did have only a single granule. Subsequently a patient with a less profound reticulocytosis was included in the survey and the variability on reticulocyte counting was considerably less. However, the number of "single dot" reticulocytes also was considerably less. A more precise delineation of the morphological criteria for identification of reticulocytes seems necessary.

The 9 year experience in morphological haematology has been a most informative experience while at times also frustrating. It has been possible to identify problem areas in morphological haematology which are sometimes due to significant differences of opinion. Hopefully the gap between these differences can be bridged. We have verified the extent of inter-laboratory variation. There is no doubt that there have been problems with the blood smears. These difficulties are inherent in the methods for smear preparation but do not invalidate certain studies carried out on these data.

V. Impressions and Conclusions

Since the first CAP haemoglobin survey in 1962, the surveys in haematology have expanded to include about 2500 laboratories in the comprehensive haematology programme and almost 4500 in the basic surveys. A tremendously valuable data base has been generated during this time and the state of the art in haematology has been well documented (Koepke, 1974).

Inter-laboratory trials have a rather unique ability to discover problem areas in laboratory medicine. The magnitude of the problem as well as potential solutions can be sorted out by careful examination of the survey data. Laboratory performance in the United States has definitely improved and has been documented.

A number of reasons for the improvement are evident. In the case of haemoglobinometry, poorer methods were abandoned and a CAP certified standard has been made widely available to all laboratories. In the measurement of the prothrombin time, inferior thromboplastins were identified and subsequently disappeared from the market. The advent of a reliable coagulation instrument has had a great impact on the performance of this test.

However, in addition to pointing out that methods and/or reagents are poor, the role of the surveys to direct laboratory workers to better methods should not be forgotten. The experience with fibrinogen determinations is an illustration of this phenomenon. No significant improvements had been found in most of the standard fibrinogen methods even though these are glaringly apparent in each survey. However, with the advent of a simple, precise method a rather striking change has been seen, and many laboratories have promptly switched to this better method.

The message to laboratory medicine is that we must continue to develop better reagents and adequate standards and if necessary introduce new methods when the presently available ones seem to be incapable of being improved. ("One cannot make a silk purse out of a sow's ear"!)

When improvement occurs it should not be pursued blindly but rather when it approaches the criterion of being "medically useful" our efforts should be redirected to other problem areas so that finally all of the procedures which we offer will be of the highest quality compatible with good patient care.

REFERENCES

Brown, D., Copeland, B. E. and Hoffman, R. G. (1963). "1962 National Hemoglobin Survey". College of American Pathologists, Chicago.

CAP Survey Manual (1974). College of American Pathologists, Chicago.

Eilers, R. J. (1965). *Biblthca haemat.* 21, 66–74.

Keitges, P. W. and Koepke, J. A. (1971). *Am. J. clin. Path.* 55, 291–301.

Koepke, J. A. (1966). "Report on Hematology, Blood Banking and Clinical Microscopy, 1965 National Comprehensive Laboratory Survey". College of American Pathologists, Chicago.

Koepke, J. A. (1970). *Am. J. clin. Path.* 54, 502–507.

Koepke, J. A. (1973). "Summing Up". 3, 5–7. College of American Pathologists, Chicago.

Koepke, J. A. (1974). *In* "Hematology Laboratory Medicine" (E. A. Stiene, ed.) pp. 79–83. Symposia Specialists, Miami.

Koepke, J. A. (1975). *Am. J. clin. Path.* (63, 990–994.)

Koepke, J. A. and Eilers, R. J. (1966). *Biblthca haemat.* 27, 137–144.

Koepke, J. A. *et al.* (1975). *Am. J. clin. Path.* 63, (984–989.)

Mathy, K. A. and Koepke, J. A. (1974). *Am. J. clin. Path.* 61, 947–958.

Miale, J. B. and LaFond, D. J. (1967). *Am. J. clin. Path.* 47, 40–59.

Rümke, C. L. (1960). *Triangle* 4, 154–158.

5. Inter-Laboratory Trials: Surveys in France

ALAIN F. GOGUEL

Central Service of Haematology, Ambroise Paré Hospital, Boulogne, France

I. INTRODUCTION

Inter-laboratory surveys in haematology have been organized in France since 1970, and 628 laboratories were included in the 1974 survey. This number constitutes approximately 1/6 of the total number of French laboratories. The survey was administered by the Haematology Department of the Faculté de Médecine, Paris Ouest, and the Federation des Syndicats de Médecines Biologistes. Combination of these two groups allowed us to include both public and private laboratories. Results from the private laboratories remain anonymous, identified only by a code. In this chapter a description is given of data treatment and information derived from the data.

II. VALIDATION AND ELIMINATION OF ABERRANT RESULTS

Firstly it is necessary to eliminate aberrant results which could cause bias in calculating means and standard deviation (s.d.).

All the standardized sheets are carefully read, completed and obvious transcription errors corrected. If necessary, they are returned to the participants for additional information. At this stage certain errors in data transcription can be recognized readily and corrected, using a histogram design. Using a computer program, validation tests are performed for each technical group. The "technical groups" are groups of participants using the same techniques; with each group verification is required for different associations: thus, for example, the Technicon method cannot be associated with a manual or semi-automated dilutor. For the numerical values, results must lie between two limits for each parameter. All values beyond these two points are printed but not treated. In the first computer run a determination is made of the "rough" mean and s.d. for each set of data, the set being defined by the sample, the test parameter, the group of techniques and the reagents used in the test. After excluding any data greater or less than 2 s.d. from the mean a new mean and s.d. are computed and are then used as references in the subsequent analysis. A result can be completely rejected; accepted but not used for the calculation of reference values; or accepted and used for the calculation of the reference values.

When each laboratory furnishes several values for the determination of the same parameter on the same sample, it is possible to use statistical tests on both intra- and inter-laboratory reproducibility (AFNOR, 1970). For testing intra-laboratory reproducibility both variances (Cochran test) and ranges are considered. However, the former is a more rigid criterion, so that groups of data with large variance are likely to be rejected, irrespective of the mean. In the evaluations which we have made, the variance test would have eliminated six times more results than tests on the mean.

Inter-laboratory reproducibility can be tested either by the relative position of results (or means of the results) from different laboratories (Dixon test) or the position of results (or means) in relation to the range of mean ± 2 s.d. This last test is very lax for a small pool (20–40 results), eliminating only transcription or punching errors which have resulted in a tenfold difference in value; however, it becomes more strict and may even be too rigid for larger pools with several hundred very close results. But this test is the only one that can be used when each laboratory gives only one result for each parameter.

It should be noted that these two groups of statistical tests will, as a rule, eliminate only a part of the data from any one laboratory (Goguel *et al.*, 1972).

III. Information from the Inter-laboratory Surveys

During the past 5 years three types of information have been obtained from the surveys.

A. INFORMATION RELATING TO ASSESSMENT OF A PARTICIPATING
LABORATORY

For each parameter and each sample, comparison of results is made with the mean of accepted values. The difference is expressed in standard deviation units: one or two asterisks are printed to indicate differences greater than 1 s.d. or 2 s.d. This comparison is made in two systems, namely (1) the technical group and (2) the reference group, which uses a standard technique which has a narrow coefficient of variation.

If two determinations of the same parameter are made on two samples, we give the position of the laboratory in a graphic representation (Youden graph) of the position of the different data of the same technical group. This position is expressed in polar co-ordinates by reference to the central point $(m1, m2)$ of the cluster, so it is easy to see the kind of error, whether systematic or random.

B. INFORMATION RELATING TO ASSESSMENT OF TEST METHODS AND MATERIALS

For each technical group, data for the same parameter are compared numerically, particularly for coefficients of variation; and graphically by means of histograms. For the parameters that have been determined on two samples, the correlation coefficient and the Youden representation are studied.

For a Youden graph, the points are transposed so that the middle of the cluster is in the middle of the graph. For each parameter the different "technical groups" are represented using the same scale, which is given by the greatest s.d. in these technical groups. Thus for the different technical groups a comparison can be made between the means (accuracy), the coefficients of variation (precision), and the coefficients of correlation and the form of the cluster (systematic error).

For the PCV (haematocrit), the smallest coefficient of variation (c.v.) in 1970 and 1971 was for the microcentrifugation technique and the biggest for the Technicon SMA 4A/7A conductivity method. With time, the situation has changed, and the smallest coefficient of variation in the last surveys was for the Coulter S although the biggest was still for the SMA 4A/7A group (Coulter S: c.v. = 2·8%; SMA: c.v. = 12·0%) (Fig. 1).

For leucocyte (WBC) determinations (Figs 2 and 3), it is interesting to look at the graph of semi-automated techniques which shows some systematic error $(R = 0·52)$ and to try to determine from which group it comes. This systematic error is present both in the "manual dilution" group $(R = 0·76)$ and in the groups with automated dilutors: Dade $(R = 0·59$; c.v. 1 = 14·3; c.v. 2 = 10·7%), Coulter $(R = 0·48$;

c.v. $1 = 17.7\%$ c.v. $2 = 12.5\%$), and Fischer ($R = 0.24$; c.v. $1 = 8.8\%$; c.v. $2 = 7.2\%$). The coefficient of variation was smaller for this last dilutor for both samples A1 and A2. However, the information is limited, and the differences could be at least partially due to some factor not connected with the dilutor, such as the type of counter used, the

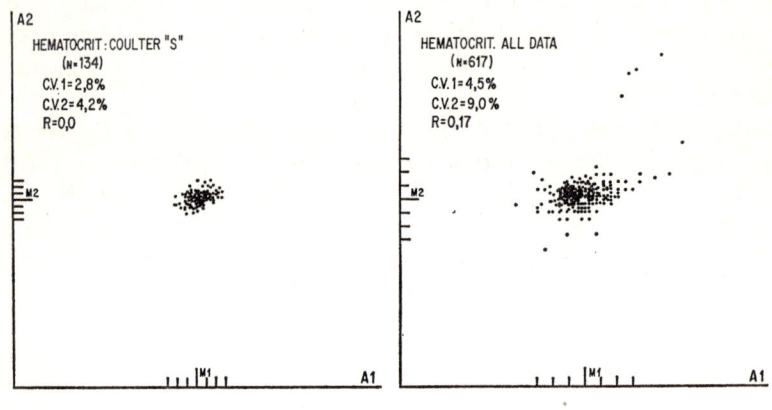

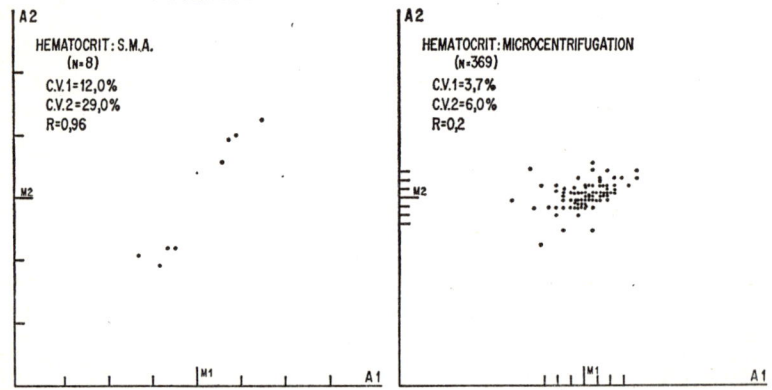

Fig. 1. Haematocrit (PCV) data on Youden plot; Etalonorme 1974 A from different technical groups (same scale).

threshold, etc. Nevertheless, there is a large systematic error in the WBC determination in the different technical groups (but not in the microscopic technique group) and this must be kept in mind if the accuracy of this determination is to be improved.

For the haemoglobin determination, the smallest coefficient of

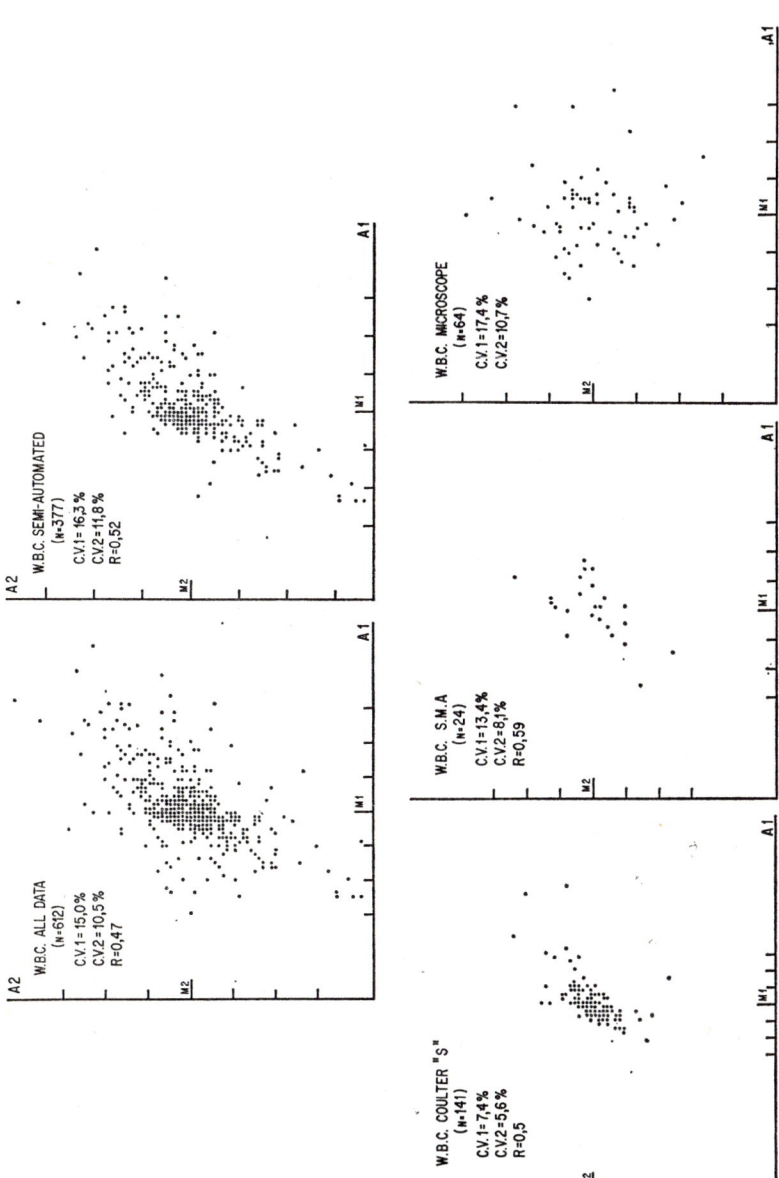

FIG. 2. WBC data on Youden plot; Etalonorme 1974 A from the main technical groups (same scale).

variation was found for the Coulter S group (c.v. $= 2{\cdot}1\%$) and this has decreased progressively, as it was between $3{\cdot}9$ and $4{\cdot}9\%$ for this same group in 1971. For the semi-automated technique in the 1974 survey the c.v. was $4{\cdot}5\%$.

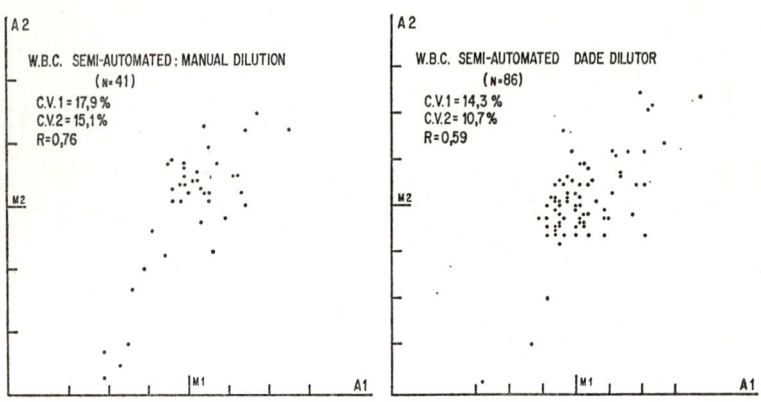

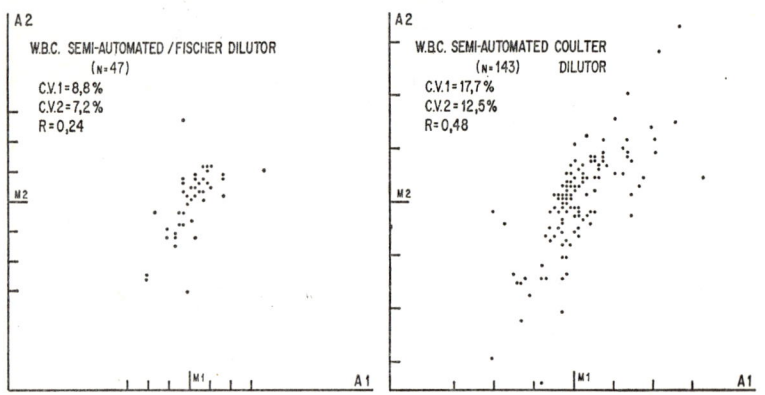

FIG. 3. WBC data on Youden plot; Etalonorme 1974 A from groups defined by the kind of diluter (same scale).

In the case of the prothrombin time, comparisons of data have been made with different thromboplastin reagents and for the bigger groups a comparison has been made of tilt tube and semi-automated techniques (Tables I and II). We wondered if it is useful to have standard "lyophylized" normal plasma in place of "fresh" normal plasma. For all but one thromboplastin, data show a smaller coefficient

TABLE I. Automation of prothrombin time determination (October 1973). Inter-laboratories survey Etalonorme 27, abnormal lyophylized plasma T.B.; Time (s.).

Thrombo- plastins	Tilt tube		Semi-automated			Fibrometer			Electromagnetic bath		
	n	m (s)	c.v. (%)	n	m (s)	c.v. (%)	n	m (s)	c.v. (%)	n	m (s)	c.v. (%)
Dade	5	19·7	7·5	5	21·2	12·6	—	—	—	—	—	—
Geigy	37	40·8	17·8	10	47·1	21·0	—	—	—	—	—	—
Merieux	44	23·4	14·1	70	24·6	12·9	48	25·5	9·7	21	22·3	15·7
Ortho	11	23·7	13·5	10	24·5	17·1	—	—	—	—	—	—
Precibio	12	18·7	6·4	31	18·6	12·4	10	19·4	16·8	13	18·3	7·0
Stago Techn.	30	22·2	13·7	30	22·0	13·0	7	22·4	9·7	21	21·3	12·3
Bio.	49	16·4	8·2	38	16·8	8·1	15	16·9	9·2	15	16·4	5·1

TABLE II. Automation of prothrombin time determination, inter-laboratories survey Etalonorme 1974 A, abnormal plasma Verify II, Time (s.)

Thrombo- plastins	Tilt tube		Semi-automated			Fibrometer			Electromagnetic bath		
	n	m (s)	c.v. (%)	n	m (s)	c.v. (%)	n	m (s)	c.v. (%)	n	m (s)	c.v. (%)
Biolyon	21	28·6	23·8	5	25·9	8·0	4	26·6	5·6	—	—	—
Dade	8	26·7	23·6	7	28·7	20·1	—	—	—	—	—	—
Geigy	37	39·4	23·4	15	44·9	9·9	7	47·1	5·0	8	43·1	11·9
Merieux	71	27·0	17·0	86	27·2	15·9	57	27·2	14·4	29	27·1	18·8
Ortho	19	24·7	10·7	15	27·1	35·2	9	25·8	11·8	6	29·1	52·4
Precibio	28	26·6	17·4	30	30·5	24·3	17	29·5	11·5	13	31·8	33·6
Stago Techn.	41	30·4	22·5	43	30·4	15·1	13	30·1	10·8	30	30·6	16·7
Bio.	67	28·4	19·4	42	30·2	20·0	12	27·6	14·4	30	31·2	20·7

of variation for "fresh" normal plasma than for "lyophylized" standard normal plasma ("Verify normal") (Table III). This result has yet to be confirmed with different preparations of lyophylized plasma.

Another question is whether there is any valid technical reason for the choice of one or other way of expressing the quick time (prothrombin time). In all groups of data, the biggest coefficient of variation

TABLE III. Prothrombin time determination, thromboplastin "Merieux" versus different plasmas

	Survey of November 1973			Survey of June 1974		
	n	m	c.v.	n	m	c.v.
Normal fresh plasma: time	116	12·6s	6·4%	171	12·5s	6·5%
Lyophylized standard plasma: time (Verify normal)	—	—	—	168	12·7s	8·2%
	E 27 abnormal plasma: technique biologique			A 4 abnormal Verify II precibio		
Time (s.)	114	24·1s	13·5%	167	27·2s	16·9%
Activity (%)	114	25·9%	23·6%	168	21·1%	25·3%
Clotting ratio/normal fresh plasma	116	1·91	13·6%	168	2·18	17·0%
Clotting ratio/Verify normal	—	—	—	165	2·12	16·0%

was for expression in percentage of activity; the smallest coefficient of variation was for the clotting ratio, either with the time of fresh plasma or the time of standard lyophylized normal plasma as denominator. The coefficients of variation of time measurement were very similar to those of clotting ratio (Tables IV and V).

TABLE IV. Prothrombin time: thromboplastin "Merieux" versus Etalonorme 27 lyophylized plasma

Coefficient of variation in different formulations of data

	n	As time (%)	As percentage (%)	As clotting ratio E 27/fresh control (%)
Tilt tube	44	14·1	20·9	6·2
Fibrometer	48	9·7	21·4	5·7
Electromagnetic bath	21	15·7	25·8	7·3
All methods	114	13·5	23·6	6·4

TABLE V. Prothrombin time: thromboplastin "Merieux" versus Verify II (A 4)

Coefficient of variation in different formulations of data

	n	As time (%)	As percentage (%)	As clotting ratios A 4/ fresh control (%)	As clotting ratios A 4/ Verify normal (%)
Tilt tube	71	16·9	27·0	17·6	19·9
Fibrometer	57	14·4	20·6	13·9	11·9
Electromagnetic bath	29	18·8	29·0	19·4	18·8
All methods	168	17·0	25·3	17·0	16·0

C. INFORMATION CONCERNING ORGANIZATION OF THE SURVEYS

In each survey we have included a study of reagents used by the participants, stability of the sample with time, and effects on these samples of condition of transport and storage. In some surveys we have

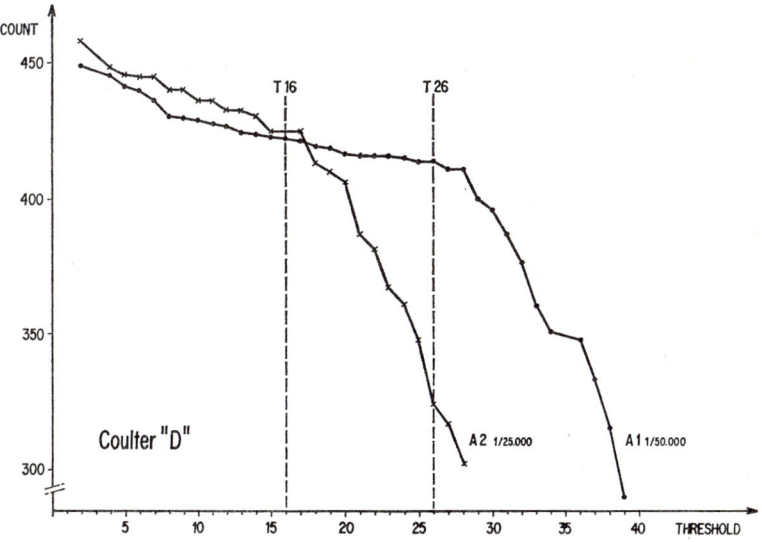

FIG. 4. Variations in RBC with variation in the threshold level of the counter. Inter-laboratory survey Etalonorme 1974 A. Count is a function of the threshold RBC samples A1 and A2.

compared two kinds of samples used for the same parameter determination. For a valid interpretation the parameter should be nearly
at the same level in both samples. Blood suspension (A1) and pre-
diluted HiCN solutions (A5) have been used in the same survey. For
hemoglobin determinations, the coefficient of variation with blood
suspension for all technical groups was 4·7%. Surprisingly, it was
greater with the HiCN solution in both the semi-automated and the
manual technique groups.

Comparison of RBC has also been made between two samples. These
were prepared in the same fashion but they had different red cell
concentrations and one of the samples (A2) had significant anisocytosis.
These differences are reflected by discrepancy in the coefficient of
variation (c.v. of A1 = 4·7%; c.v. of A2 = 6·5%) (Table VI). One of
the mechanisms of this discrepancy could be variation in the count with
variation in the threshold level of the counter (Fig. 4).

TABLE VI. Inter-laboratories survey Etalonorme 1974 A, RBC ($\times 10^{12}/l$)

Technique	n	A 1			A 2		
		m	s	c.v.(%)	m	s	c.v.(%)
Microscope	55	5·01	0·36	7·78	2·56	0·24	9·37
Semi-automated	392	4·91	0·24	4·88	2·64	0·16	6·06
Coulter S	143	4·85	0·11	2·26	2·65	0·11	4·15
Technicon S.M.A.	23	4·97	0·18	3·62	2·69	0·20	7·43
All methods	619	4·91	0·23	4·68	2·63	0·17	6·46

REFERENCES

Association Française de Normalisation (AFNOR) (1970). "Guide pour les essais
    inter-laboratoires". N.F. X 06-041.
Goguel, A., Guerit, D., Beuzart, A. and Speth, J. (1972). In "Organisation des
    Laboratoires, Biologie Prospective, Ilème colloque de Pont-a-Mousson". L'Expan-
    sion Scientifique Française Ed., Paris.

6. Standards and Reference Preparations

S. M. LEWIS*

Royal Postgraduate Medical School, London W12 0HS, England

I. DEFINITIONS OF STANDARDS

A number of different materials which are used in standardization and quality control are frequently referred to loosely as "standards". These include international standards, international reference preparations, national standards, working standards, standard reagents or solutions, reference samples and control samples.

A primary analytical standard has been defined as a pure chemical substance used for the purpose of assaying a solution of unknown strength or for the preparation of a solution of known concentration (Radin, 1967). It should have the following criteria (Farr *et al.*, 1951):

(1) It must be a stable substance of definite composition.

(2) It must be capable of being dried in the course of preparation, preferably at 105–110°C without change in composition.

* In receipt of grant from British Department of Health and Social Security.

(3) It should have a high equivalent weight in order that weighing errors may have relatively small effect.

(4) It must be a substance that can be accurately analysed.

(5) Desired reactions should occur according to a single well-defined, rapid and essentially complete process.

(6) Its purity must be assured through well-defined qualitative tests of known sensitivity or through preparation by a method that has been consistently shown to yield a pure product and by storage under conditions in which the product is entirely stable.

Some of the materials used in haematology are chemical, and for their standardization they should conform to all the criteria. Others are biological and for these a more limited control may suffice. The World Health Organization (1968) has prescribed criteria for biological standardization. These require that an international standard shall be a preparation of precisely defined content, with long term stability and to which an international unit has been assigned on the basis of an extensive international collaborative study. In this category are included the majority of antibodies and antibiotics, hormones, vitamins and some enzymes. When a substance is not entirely suitable to serve as an international standard, perhaps because it is not sufficiently stable, it may still be possible to define its composition and purity and the potency or concentration of each batch manufactured. Material which fulfils these more limited criteria is known as an international reference preparation; an example is the hemiglobincyanide reference preparation.

When a chemical reference substance can be characterized completely by chemical and physical tests, it should no longer be necessary for material as such to be held by reference centres. This is the ultimate aim of all international standards. Thus, for example, in the case of haemoglobin, it may be possible to produce a glass filter of appropriate spectral absorbance which could be used as a constant physical reference for calibrating spectrophotometers which are used to measure the haemoglobin solutions.

In practice, international standards and reference preparations serve much the same purpose. Stocks of each are held by designated laboratories and they are used for establishing the equivalent national standards. As a rule, a limited amount is available so that it is supplied only to national reference centres; it cannot be supplied at random to laboratories throughout the world for routine use.

The standards or reference preparations which are established nationally should be identical with or strictly comparable to their international counterparts. They should be the basis for calibration of materials which are manufactured commercially or by individual laboratories for use in the test itself. The national standard is sometimes referred to as a primary standard, while the preparation used for comparison of a colour reaction is known as a working standard, or secondary standard. It is preferable to use an expression such as comparator reagent in order to distinguish this material from a true standard, albeit in some instances it might well be similar or, indeed, identical to the reference preparation.

A. STANDARDIZED REAGENTS AND REFERENCE METHODS

Standardized reagents are necessary when the reference preparation is used in a complex chemical reaction. An international standard reagent, with an assigned composition and purity, should be used in a reference method, in conjunction with an international reference preparation. Reference methods are becoming recognized as an integral part of standardization; indeed there are a number of situations where the procedure rather than a specific chemical material can be standardized. Sometimes the reference method can also be used routinely but when it is complex it may be necessary to recommend a simpler procedure for routine practice. In such cases, and particularly when the routine method is carried out in an automatic analyser, it is essential to check its accuracy and reliability by comparison with the reference method.

B. REFERENCE SAMPLES

The ultimate aim of the use of standards and standard methods is to achieve acceptable levels of accuracy and precision. Essential for this are intra-laboratory quality control and inter-laboratory proficiency assessment. Two forms of preparation are required: (1) *reference samples* of stated value for calibration of instruments, for establishing the accuracy of a new or modified method and to ensure that there is no consistent error or drift;* (2) *control test samples* of unstated value for participating laboratories to carry out tests in order to assess variance from the true value (as determined by a reference method) and from the modal values (based on the computation of results from all participants).

Control test samples are treated exactly as routine test samples and should be assayed concurrently, whereas a reference sample of known value is usually used under special test conditions. In view of the important implications of its use the reference sample should be of unquestionable reliability, preferably controlled and certified by a national authority. Minimal requirement for this is an analysis of its components in an expert laboratory by more than one method, and the

---

* This can be demonstrated by plotting the cumulative sum of differences between successive measurements and the reference value ("cusum") (Cavill and Jacobs, 1973).

certification should include a statement of mean value, limits of deviation and, where appropriate, an assurance that the stock has been calibrated against a national reference preparation.

Adequate quality control thus requires a number of different preparations, although some may be interchangeable. In such cases, if large supplies of a national reference preparation were freely available, they would serve well as a reference sample or even as a control test sample. On the other hand, a test preparation may be suitable for one aspect of quality control but not for another. Thus, for example, for inter-laboratory trials the material must be stable for at least several days when left at ambient temperature and subjected to the trauma of postal delivery, whereas material for intra-laboratory use can be gently handled and maintained at 4°C.

In this chapter standards used in haematological tests will be considered.

II. Haemoglobin Standard

ICSH published recommendations for haemoglobinometry in human blood in 1967 and proposed criteria for a reference standard of hemiglobincyanide (cyanmethaemoglobin). This consists of human blood lysate which is converted to hemiglobincyanide and dispensed as a sterile solution in 10 ml ampoules of amber glass. The haemoglobin content is measured by spectrophotometry and samples are tested for purity, stability and sterility by a number of expert laboratories nominated by the committee and sited in different parts of the world. The standard is prepared on behalf of ICSH by the Netherlands Institute of Public Health (RIV). Because it was originally thought that the material would be stable for about 1 year, a new batch is manufactured each year. More recently it has been shown that it has considerably longer stability than indicated by the expiry date stated on the label (Holtz and Van Assendelft, 1973). Supplies are provided by ICSH to national standards committees, who are expected to make them available to manufacturers and distributors in their countries to ensure that comparator reagents ("working standards") made locally conform to the international reference preparation. The World Health Organization adopted the ICSH standard as an international reference preparation in 1968. During the past few years there have been proposals for alternative methods. However, the ICSH recommendations remain virtually unchanged at the present time.

A number of standards organizations (e.g. British Standards Institute 1966) have developed national reference preparations which are based on the ICSH recommendation. This ensures control in accordance with national legislation and practice. The British Committee for Standards in Haematology controls and certifies preparations manufactured in the United Kingdom, and a similar service is provided in USA by the College of American Pathologists. The ICSH reference preparation is used by the BCSH testing laboratory to ensure that its spectrophotometers are reliable and that the material under scrutiny can be equated with the international reference preparation. As these commercial products are used routinely as a comparator reagent this ensures that haemoglobin estimations carried out in clinical laboratories throughout Britain are based on the international standard.

III. Lysed Blood Standards

The ICSH reference preparation provides a reference point for measurement of haemoglobin as hemiglobincyanide. When haemoglobin is measured as oxyhaemoglobin it is necessary to have a different comparison reagent, derived from whole blood. Also, in some automatic haemoglobinometers the measurement of haemoglobin is performed by a photoelectric colorimeter which is preset to give direct readings of haemoglobin values after adjustment of the controls against a whole blood standard. For these purposes, fresh normal adult blood can be used provided that the haemoglobin content is first determined as hemiglobincyanide by a photoelectric colorimeter or spectrophotometer. Alternatively, lysed blood standards are available commercially, but these tend to deteriorate during storage, with formation of methaemoglobin and protein denaturation; furthermore repeated opening of the bottle for subsampling results in evaporation and contamination by micro-organisms. These problems can be largely overcome by dispensing the blood into aliquot samples which are stored at $-20°C$ and used once only. This material is said to preserve better when collected into glycerol (Mather, 1966) but this is more complex to prepare and, because it is sticky and viscid, difficult to use.

Whole blood is also used as a control test sample. Whereas use of the haemoglobin reference preparation ensures reliability of instrumentation, when whole blood is used all parts of the test procedure will be checked, including dilution, which is an important source of error. For this, the blood can be lysed or non-lysed. The latter has an added advantage that it will demonstrate errors due to inadequate sample mixing. Differences between measurement of haemoglobin by different types of instrument and between measurement of whole blood and hemiglobincyanide solution by the same type of instrument in an interlaboratory trial are shown in Table I.

TABLE I. Inter-laboratory trial of haemoglobinometry

		Haemoglobin (g/dl)		Variance index*		
	No.	Mean	s.d.	Mean	s.d.	No. with index >2·0
(A) BLOOD SAMPLE						
Photoelectric colorimeters	171	9·83	0·30	0·83	0·50	10
Automated counters (e.g. Coulter S)	108	9·60	0·17	0·73	0·67	12
(B) HiCN SOLUTION						
Photoelectric colorimeters	207	11·48	0·23	0·90	1·43	27

Haemoglobin was measured (A) on a whole blood sample and (B) on a solution of hemiglobincyanide; results expressed in g/dl.

* Variance index compares the deviation of an individual result from the mean, and the s.d. determined from all results (excluding outliers). It is calculated separately for each participant from the formula $x - \bar{x}/s'$ where x = value as obtained, $\bar{x}$ = weighted mean and s' = 1 weighted s.d.

IV. RED BLOOD CELLS

The characteristics required for an adequate standard for red-cell counting (RBC) are as follows: (1) it should have reasonably long stability, (2) it can be used directly, (3) it will suspend readily and will not agglutinate, (4) its rheological constant should be similar to blood, (5) it will behave in a manner similar to red blood cells with regard to index of refraction and/or specific conductivity when taken up into red-cell diluting fluids, (6) it maintains size and shape in a way comparable to blood, and (7) it can be assayed by independent methods.

A large number of materials have been tried. They fall into two main groups—(a) artificial material, and (b) natural blood cells, either preserved or modified by fixative treatment. Artificial materials include pollens, mould spores, yeast, polystyrene latex and other plastic polymers; none of these has proved suitable. Natural blood has many advantages but a major disadvantage is its lack of stability. When blood is collected into EDTA and kept at 4°C it can be used for up to 24 h for RBC, haemoglobin, PCV and absolute values (MCV, MCH and MCHC) (Brittin et al., 1969). For RBC the blood is usually stable for several days, but PCV tends to deteriorate after 24 h (Figs 1–2). If stored at c. 20°C the red cells swell rapidly rendering the blood unsuitable for PCV and derived parameters after a few hours.

Blood collected into a preservative-anticoagulant solution is more stable. Three solutions are in common use: (1) acid-citrate-dextrose [ACD (NIH-A)], pH 6·4; dextrose 25 g, trisodium citrate 22 g, citric acid 8 g, water to 1 l, 20 vol. blood to 3 vol. solution;

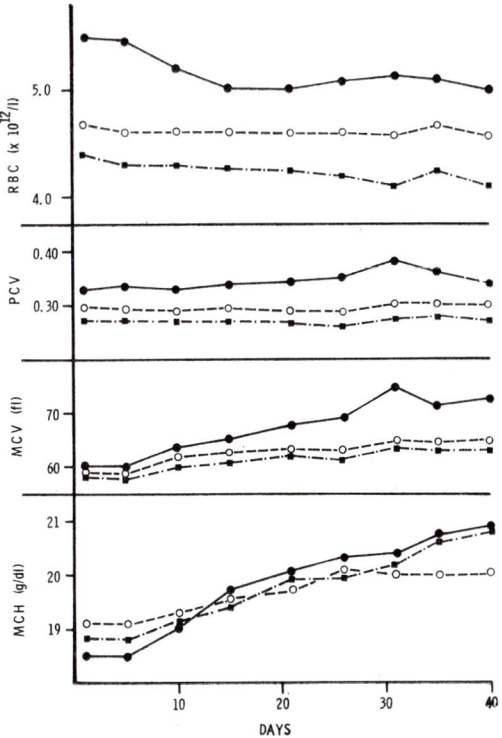

FIG. 1. Effects of anticoagulant on blood counts. The specimens were kept at 4°C and equilibrated at room temperature (*c*. 20°C) just before the tests were carried out on a Coulter S Cell Counter. ●, EDTA; ○, ACD (NIH-A); ■, Alsever's solution.

(2) citrate-phosphate-dextrose (CPD) pH 6·9: dextrose 2 g, trisodium citrate 30 g, sodium dihydrogen phosphate 0·15 g, water to 1 l, 1 vol. blood to 2 vol. solution;

(3) citrate-glucose (Alsever's solution) pH 6·1: dextrose 20·5 g, trisodium citrate 8 g, citric acid 0·55 g, sodium chloride 4·2 g, water to 1 l, 4 vol. blood to 1 vol. solution.

In these solutions RBC is maintained at a constant level for a few days at *c*. 20°C and for 3–4 weeks at 4°C. The best medium appears to be NIH-A ACD in which stability of both RBC and PCV is maintained for several weeks (Fig. 1); with some batches these parameters have remained constant for up to 12 weeks. MCV and MCH are less stable, at least when measured by certain techniques (see below).

Some workers have recommended adding inosine, adenine or ouabain to the ACD or Alsever's solution (Åkerblom et al., 1967; Haradin et al., 1969; Odake et al., 1969). These additives undoubtedly improve red-cell integrity and subsequent in vivo survival of the blood when transfused (Haradin et al., 1969; Odake et al., 1969); but it is debatable whether there is significant prolongation in the stability of

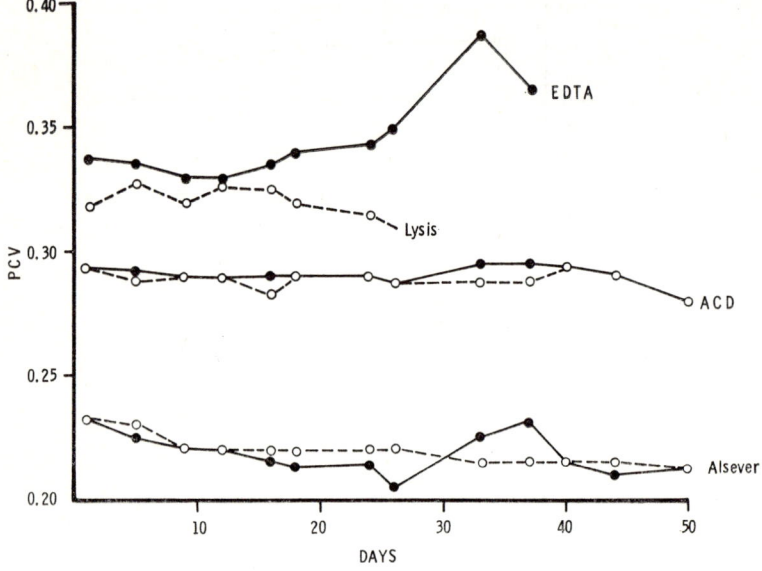

Fig. 2. Comparison of PCV measurement by microhaematocrit (○) and Coulter S(●) on stored blood in different anticoagulants.

the RBC parameters. It has been shown that when adenosine restores cellular ATP, there is no reversal of membrane lipid loss, so that sphere-disc transformation occurs with the production of a smaller sized disc than the original red cell (Haradin et al., 1969). Zucker and Brosius (1970) recommended the addition of adenine, hydrocortisone, Phenergan and Cohn's plasma fraction IV to blood collected into CPD; in this solution they found the RBC and PCV to be stable for a 30 day period at 4°C with less than 3% of specimens showing any lysis, and even then, only minimal amounts. Our own experience suggests that Alsever solution is marginally improved by the addition of inosine 3·6 g, adenine 0·24 g and tocopherol phosphate 0·1 g/l. However, the best medium appears to be ACD (NIH-A), as discussed above. It is advisable to include an antibiotic: thus e.g. 2 ampoules of

Crystamycin (Glaxo), each containing 300 mg sodium penicillin G and 500 mg streptomycin, should be added per litre of preserved blood.

The problem of using preserved blood as a reference preparation or as a control for inter-laboratory trials is complicated by the fact that different instruments respond differently to the same preparation (Figs 2 and 3) and, conversely, in some cases the same instrument gives

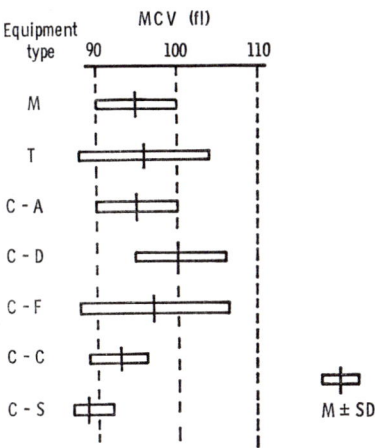

FIG. 3. Comparison of MCV measurements (or calculation) by various techniques in an inter-laboratory trial on blood preserved in ACD. Mean and 1 s.d. are shown. The instruments used were as follows (no. of participants in each group indicated): M = haemocytometer and microhaematocrit (14); C–A = Coulter A and microhaematocrit (12); C–D = Coulter D and microhaematocrit (15); C–F = Coulter F and microhaematocrit (13); C–C = Coulter F and MCV computer (15); C–S = Coulter S (46).

a different pattern of results with each preservative (Fig. 2). In evaluating the preparation it is necessary to distinguish errors of performance from faulty material, and to assess how well the material stands up to varying conditions of storage. Inter-laboratory trials are useful for this. In one such trial, blood in ACD appeared to maintain its original measurements for at least 3 weeks, as shown by mean values and s.d. (Fig. 4). The samples were kept by the participants under "routine conditions of storage", generally at 4°C. However, about 20% of the samples were found to be lysed at the time of the second test, and were not included in the analysis.

Low temperature preservation of blood in glycerol has been tried (Mather, 1966) and although this suspension is relatively stable for several months, progressive lysis and a proportional fall occurs in RBC and PCV.

FIXED BLOOD CELLS

Blood can be permanently stabilized by fixation. Unlike the process of preservation described above, fixed blood undergoes a physico-chemical change. Formaldehyde (Benedek, 1966), tanning (Orthey *et al.*, 1965) and glutaraldehyde (Lewis and Burgess, 1966; Lewis, 1972) have been used as has a fixative solution of acetic acid with sodium sulphate (Torlontano and Tata, 1971). Glutaraldehyde-fixed cells shrink in size; at first this is considerable, and then the process continues slowly for 3–4 days. Thereafter changes are virtually undetectable and

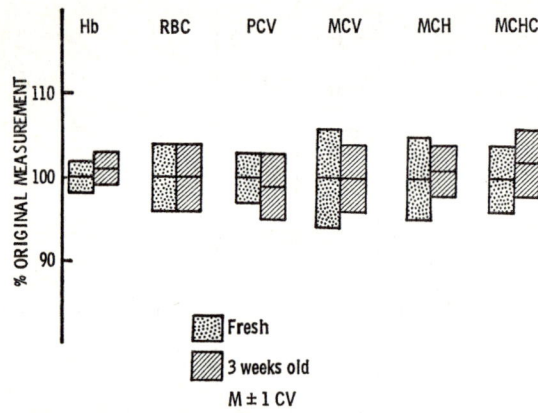

Fig. 4. Effects of aging on blood preserved in ACD, tested in an inter-laboratory trial by 140 participants using various techniques.

the preparation gives consistently reproducible measurement of cell numbers and cell size distribution for months and even years. There are, however, a number of disadvantages with fixed cell preparations. They have greater viscosity than normal blood and a different relative flow rate (Ham *et al.*, 1968). In electric counters this will influence the number of particles passing the counting area in a given time, with an obvious effect on the count. In cell counters based on the Coulter system, as cells pass through the orifice their volume directly influences pulse amplitude. Natural cells deform as they pass through the orifice and this affects the characteristics of the pulses (Thom, 1972), whereas the fixed cells are inflexible, and this results in a unimodal distribution of pulses (Lewis, 1972). Furthermore, there is a tendency for glutaraldehyde-fixed cells to clump and to adhere to the walls of the plastic containers, so that unless the sample is suspended by vigorous shaking, there may be non-homogeneity and loss of precision with replicate tests (Fig. 5). Another disadvantage is that in measuring

PCV by haematocrit the rigidity of the fixed cells results in an erroneously high PCV, although the results are consistently reproducible when the test is carried out in a precise manner. Thus, glutaraldehyde-fixed cells have some value in quality control, but they are more useful as a reference preparation for instrument calibration.

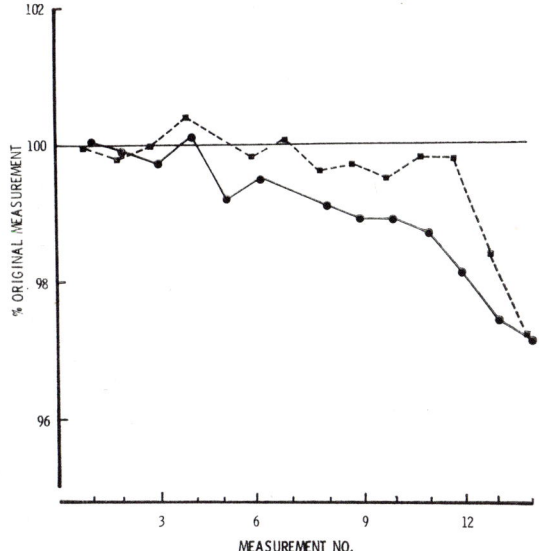

Fig. 5. Sequential measurement of RBC on fresh blood (■) and glutaraldehyde-fixed blood (●). Each measurement was the mean of the second and third readings of a dilution.

Cells treated with acetic-acid-sodium sulphate procedure are said to behave more like natural red cells with no change in size and in size distribution for up to 12 months (Torlontano and Tata, 1971).

ANIMAL BLOOD

Availability of adequate supplies of blood for use in quality control has become a major problem in many countries. For most purposes animal blood is a suitable source, and donkey blood is especially useful as the red cells have an MCV of 50 fl and the same relative size distribution as human blood (Fig. 6). Thus, an instrument calibrated with this reference is likely to be correctly set for counting most of the cells which occur in human blood in disease as well as in health (Fig. 7).

Establishing the cell count in a primary standard is a major problem, as the measurement should not be carried out on an electronic counter with a relatively arbitrary calibration. Theoretically, the only acceptable method to obtain the true count is by direct counting of the cells

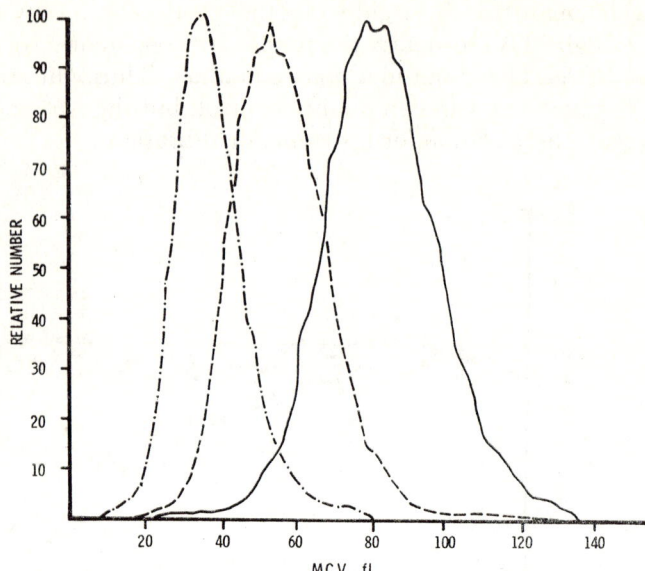

Fig. 6. Frequency distribution of cell volume for normal human (———) and donkey (— — — —) red cells, and for donkey cells after glutaraldehyde fixation (— · — ·).

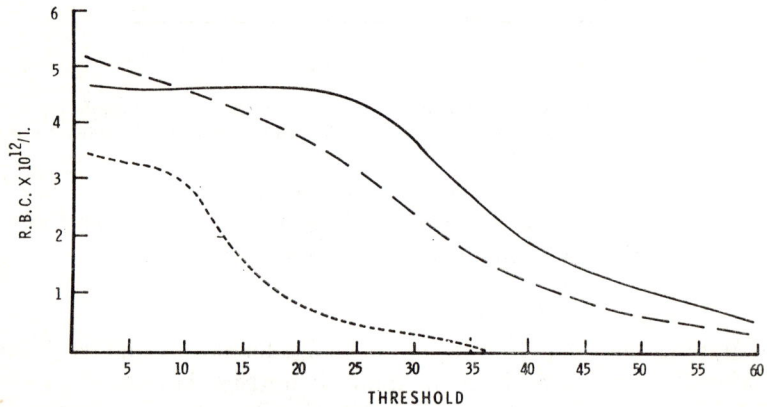

Fig. 7. Effects on RBC of various setting of the Coulter Counter (model F_n) with normal human blood, donkey blood and blood from a patient with thalassaemia. When adjusted by normal blood the counter may not be correctly set for counting abnormal blood.

in diluted suspension in a haemocytometer counting chamber. This well established procedure can be made sufficiently accurate by using a chamber which conforms to precise specifications (e.g. those of the British Standards Institution, BS 748:1963), and by carrying out a sufficiently large number of replicate counts to minimize errors due to random cell distribution. It is, of course, a time-consuming technique but by photographing each area the count can be made leisurely, methodically and without bias (Dacie and Lewis, 1968).

There is not yet any internationally agreed standard or reference preparation of red cell, nor any certification schemes similar to those for haemoglobin. A number of commercial materials are available which are intended primarily for calibrating electronic counters. Preliminary studies have shown that none of these meet all requirements for an ideal reference standard (Helleman, 1972). A major problem with all such preparations is that their deterioration on storage results in a greater variability than the precision of the instruments which they are intended to control.

V. Leucocytes

Blood in EDTA kept at 4°C is stable for leucocyte counts for 24 h. When collected into ACD the leucocytes are stable for up to a week. Acetic acid-sodium sulphate is said to preserve leucocytes for a considerable time (Torlontano and Tata, 1972), and an artificial suspension of *Candida albicans* has also been used (Lappin and Sanderson, 1970). The most stable material consists of glutaraldehyde-fixed cells. By suspending an appropriate volume in ACD-preserved human (or animal) blood, the preparation simulated whole blood for leucocyte counting by normal routine procedure. This has been found to be suitable for use with various counters in inter-laboratory trials (Fig. 8). The wide s.d. suggests that in some cases the tests were carried out with instruments not used at correct calibration settings. In this study turkey erythrocytes which are nucleated cells were used. These have a disadvantage in that they are reliable for calibrating Coulter counters only when one commercial lysing agent, Zaponin,* is used for subsequent leucocyte counts, as this does not alter the size of the unfixed leucocytes or the fixed cells whereas other lysing agents affect the unfixed leucocytes, but not the fixed cells, to a variable extent (Lewis, 1972). More recent studies have suggested that fixed human erythrocytes (nonnucleated cells) are more satisfactory as a leucocyte substitute than turkey cells, but the problem of the lytic agent remains.

VI. Platelets

No suitable material is yet available. Suspension of *Micrococcus roseus* has been used (Mayne and Carville, 1969), as have glutaraldehyde fixed platelets obtained by concentration from whole blood collected

* Coulter Electronics Ltd.

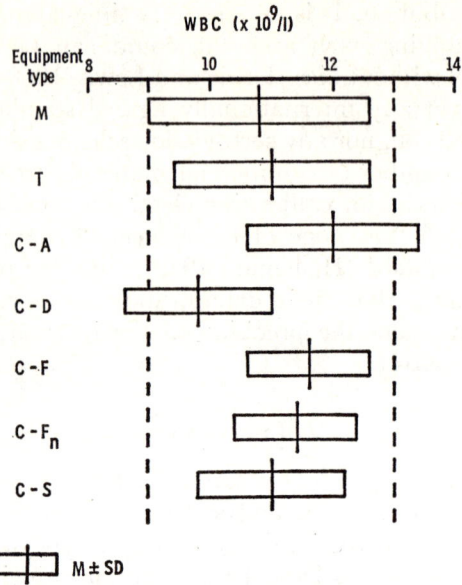

Fig. 8. Comparison of leucocyte count by various techniques in an inter-laboratory trial on a glutaraldehyde-fixed preparation. Mean and 1 s.d. are shown. See Fig. 3 for explanation of instrument code.

into ACD. The latter preparation gives an excessively wide variance when used in inter-laboratory trials (Table II). This appears to be due both to the technical difficulties of platelet counting and to the unsatisfactory nature of the test material.

TABLE II. Results of platelet counts on two preparations containing glutaraldehyde-fixed platelets

Sample	No. of participants	Mean platelet count $(\times 10^9/l)$	s.d.	Range $(\times 10^9/l)$
A	198	200·1	59·5	16–500
B	200	135·7	51·6	15–511
Ratio of B/A		0·70	0·21	0·3–6·3

VII. Serum Iron

ICSH (1971) has recommended a reference method for measurement of serum iron in human blood and specifications for a standard reference preparation. These were developed in an extensive study which was carried out by the Expert Panel (Izak and Lewis, 1972). The estimation of iron is by a spectrophotometric procedure in which the colour reaction of serum treated with a chromogen solution of bathophenanthroline sulphate is compared with that developed by a standard solution of iron. This iron solution is a chemically-pure inorganic preparation consisting of a weighed amount of electrolytic iron wire dissolved in HCl. For the standard reference preparation, on the other hand, sterile pooled haemoglobin-free serum is obtained from normal (human) subjects and divided into two lots. To one is added a known amount of iron from a standard solution and both lots are then lyophilized to dryness. The specifications of these paired preparations include (1) instruction concerning the volume of iron-free water required to reconstitute them to their original volume, (2) their iron concentration as determined by designated reference laboratories using the reference method, and (3) expiration date when stored at 4°C.

VIII. Erythrocyte Sedimentation Rate

ICSH (1973) has established a reference method for the ESR test which includes specifications for the Westergren tube. Several national standards organizations have adopted these criteria, including the British Standards Institution which has published the specifications as BS 2554: 1968, and in the USA it has been adopted by the National Committee for Clinical Laboratory Standards. There is, however, still need for a control procedure to ensure not only that the commercially manufactured tubes conform to the specifications, but that when the test is carried out in accordance with the recommended method it gives consistently reliable results. In this particular test there is no defined measure of accuracy, so that, at the present time, only reproducibility (i.e. precision) and clinical significance can be used as criteria of acceptability. There is need for a reference material, but its form and composition defy solution at this time. It might be some artificial preparation with defined rheological qualities, or possibly blood obtained freshly from a donor known to be normal—a walking standard!

IX. Other Standards

Wheresoever there is need for standardization and quality control, there is also need for standards. This is the case with many haematological tests, and ICSH Expert Panels, aware of the problem, are carrying out extensive in some of these areas. Inter alia: (1) reference thromboplastins and reference plasmas for standardizing the prothrombin time test used in oral anticoagulant control (see p. 153); (2) reference samples of abnormal haemoglobins and of Hb-F and Hb-A_2; (3) standardization of antiglobulin reagents, especially anti-IgG and anti-complement; (4) red-cell enzymes; (5) the specimen container itself (see p. 211).

Adoption of standardized methods does not imply an immutable state without further progress. Methods must be continually re-evaluated and new techniques must be developed as horizons alter. The standards, too, may require amendment as scientific progress provides means for more critical measurement and control. This review has demonstrated some of the achievements of standardization in haematology, but it has also indicated how limited these achievements have been in comparison to the needs, and how long is the road ahead.

REFERENCES

Åkerblom, O., De Verdier, C. H., Finnson, M., Garby, L., Hogman, C. F. and Johansson, S. G. O. (1967). *Transfusion, Philad.* **7**, 1–9.

Benedek, E. (1966). *Bibl. haemat., Basel* **24**, 67–70.

British Standards Institution (1966). "Cyanmethaemoglobin Solution for Photometric Haemoglobinometry. BS 3985".

Brittin, G. M., Brecher, G. Johnson, C. A. and Elashoff, R. M. (1969). *Am. J. clin. Path.* **52**, 690–694.

Cavill, I. and Jacobs, A. (1973). *ACP Broadsheet No. 75.* Association of Clinical Pathologists, London.

Dacie, J. V. and Lewis, S. M. (1968). "Practical Haematology" 4th edn, pp. 24–29. Churchill, London.

Farr, H. V., Butler, A. Q. and Tuthill, S. M. (1951). *Analyt. Chem.* **23**, 1534.

Ham, T. H., Dunn, R. F., Sayre, R. W. and Murphy, J. R. (1968). *Blood* **32**, 847–861.

Haradin, A. R., Weed, R. I. and Reed, C. F. (1969). *Transfusion, Philad.* **9**, 229–327.

Helleman, P. W. (1972). *In* "Modern Concepts in Hematology" (G. Izak and S. M. Lewis, eds) pp. 235–239. Academic Press, New York and London.

Holtz, A. H. and Van Assendelft, O. W. (1973). *In* Proceedings of 2nd Meeting of the European and African Division of the International Society of Hematology, Prague, p. 27.

International Committee for Standardization in Hematology (1967). *Br. J. Haemat.* **13** Suppl., 71–75.

International Committee for Standardization in Hematology (1971). *Br. J. Haemat.* **20**, 451–453; *Am. J. clin. Path.* **56**, 543–545.

International Committee for Standardization in Hematology (1973). *Br. J. Haemat.* **24**, 671–673.

Izak, G. and Lewis, S. M. (Eds) (1972). "Modern Concepts in Hematology", pp. 69–160. Academic Press, New York and London.

Lappin, T. R. J. and Sanderson, F. M. (1970). *J. clin. Path.* **23**, 65–67.

Lewis, S. M. (1972). *In* "Modern Concepts in Hematology" (G. Izak and S. M. Lewis, eds) pp. 217–229. Academic Press, New York and London.

Lewis, S. M. and Burgess, B. J. (1966). *Lab. Pract.* **15**, 305–306.

Mather, A. (1966). *Clinica chim. Acta* **13**, 141–146.

Mayne, E. E. and Carville, J. M. (1969). *J. clin. Path.* **22**, 107–109.

Odake, K., Bishop, C., Warner, W. and Ambrus, J. L. (1969). *Vox Sang., Basel* **17**, 375–392.
Orthey, G. F., Traynor, J. E. and Ingram, M. (1965). *Bibl. haemat., Basel* **12**, 14–24.
Radin, N. (1967). *Clin. Chem.* **13**, 55–76.
Thom, R. (1972). *In* "Modern Concepts in Hematology" (G. Izak and S. M. Lewis, eds) pp. 191–200. Academic Press, New York and London.
Torlontano, G. and Tata, A. (1971). *Acta haemat.* **45**, 325–329.
Torlontano, G. and Tata, A. (1972). *In* "Modern Concepts in Hematology" (G. Izak and S. M. Lewis, eds) pp. 230–234. Academic Press, New York and London.
World Health Organization (1968). Technical Report Series No. 384.
Zucker, S. and Brosius, E. (1970). *Am. J. clin. Path.* **53**, 474–480.

7. Problems of the Red-Cell Volume

P. CROSLAND TAYLOR

School of Pathology, Middlesex Hospital Medical School,
London W1, England

I. INTRODUCTION

There are three principle methods of measuring the red-cell volume of a sample of blood. First described by Stewart (1899), they are (a) centrifugation, (b) indicator studies (i.e. chemical, dye or isotope dilution), and (c) measurements of conductivity. The different methods give approximately similar values. However, on theoretical grounds, they should not give precisely the same values and this is found in practice. With some blood samples the differences are much greater than with others, and with increasing use of conductivity measurements, the discrepancies between different methods have become both apparent and important.

The importance of the differences between methods depends on technical precision, the range of values in health and disease and the variations which may occur over a period of time in an individual who remains healthy.

The greatest experience of the use of cell volume to differentiate health from disease has been gained from centrifugal measurements. The range of values, both in health and disease, is greatest when the measurement is taken in isolation without reference to red-cell count or haemoglobin. In healthy adults it is 0·35–0·54 (Dacie and Lewis, 1968) and in disease the range extends from 0·08–0·80.

When expressed in relation to the red-cell count as the mean cell volume (MCV), the range is reported as 76–96 fl in health and 50–140 fl in disease. When related to the haemoglobin, expressed as the mean corpuscular haemoglobin concentration (MCHC), the corresponding ranges are 30–35% and 25–38%. The above values, based on cell volumes obtained by centrifugation, indicate that differences in method will be most influential for the MCHC and least for the packed cell volume (PCV).

II. CENTRIFUGATION

When a column of blood is centrifuged in a tube of uniform bore, the proportion of cells to the whole, known as the packed cell volume (PCV) or haematocrit,* excludes the platelet layer, but includes some or all of the leucocytes. The values obtained are increased by the fraction of the plasma trapped in the cell column, but decreased if the cells can be compressed by exposure to high gravitational forces (g). Using Wintrobe haematocrit tubes spun for 30 min at 1500 g, the trapped plasma amounts to 3–4% if the blood is from a normal healthy adult. Fifty-five minutes centrifugation reduces trapping to 2–3% (Chaplin and Mollison, 1952). At higher g trapping is less (Garby and Vuille, 1961), with a correspondingly lower PCV (Stewart, 1966). In certain diseases, such as iron-deficiency anaemia, and sickle-cell anaemia, or in the blood from certain animals, the proportion of trapped plasma is greater. These observations account for an accentuation of the rate of fall of the MCHC and a reduction of the rate of fall of the MCV during the development of iron-deficiency anaemia, when cell volume is measured by centrifugation. Conversely, Ponder and Saslow (1930) and Sirs (1968) have suggested that distortion and possibly compression of cells by centrifugation give falsely low values.

Other factors which influence accuracy and precision of PCV measurement include oxygenation or deoxygenation of the blood sample which will decrease or increase the value respectively, and variation in bore of the haematocrit tube. Under constant physical conditions, reproducibility of PCV on duplicate samples should be approximately 1%, for MCV and MCHC the variance is further increased by variance of the measurements of Hb and RBC.

There are no universally accepted standards for centrifugation nor are conditions always specified when reference is made to the PCV.

---

* Contrary to current convention, it would be more correct to talk of the packed red-cell volume or PRCV. The term "haematocrit" literally means "blood separation".

Commonly, centrifugation for 30 min 1500 g is used with Wintrobe tubes, and 6–12 000 g for 5 min for capillary tubes (see also p. 106).

III. CHEMICAL, DYE AND ISOTOPE DILUTION

These methods depend on the dilution of a substance miscible only with the suspending medium. They can conveniently be subdivided according to the sensitivity with which the added material may be measured.

An example of a crude method is the measurement of glucose or albumin in the suspending medium by chemical means before and after the addition of known volumes of fluid containing these substances. Examples of more sensitive methods are the measurement of plasma concentration of a dye such as Evans blue which binds to albumin, or the pigment haemoglobin, before and after the addition of a known small volume to a measured sample of blood. The advantage of the latter method is the greater precision of direct spectrophotometry in comparing the concentrations of pigment in two otherwise identical solutions.

The most sensitive methods are based on the addition of small quantities of radioactive material such as radioiodinated human serum albumin (RIHSA). The advantages are as follows. (i) Comparison of the radioactivity of two solutions is as precise as colour comparison by spectrophotometry. (ii) It is unnecessary to add a measured volume of RIHSA to a known volume of blood provided that the quantity added is too small to make any significant change in the cell volume. (iii) The ability to measure radioactivity in optically different solutions (i.e. whole blood or plasma) means that after centrifugation the proportion of trapped plasma may also be measured. (iv) The method can be used in conjunction with other measurements as the trace of RIHSA will not interfere with the sample.

IV. ELECTRICAL CONDUCTIVITY

Estimation of cell volume from observation on the interference of the electrical conductivity of the suspending medium can be made by one of two methods. The first is based on Maxwell's equation for homogenous suspension of non-polarizable spheres, modified by Velick and Gorin (1940) for non-conducting non-spherical particles:

$$p = \frac{\dfrac{K}{K_1} - 1}{\dfrac{K}{K_1} - 1 + F}$$

where p = fraction occupied by cells, K and K_1 are the specific con-
ductivities of the suspension and suspending medium and F is a shape
factor which was found by Hirsch *et al.* (1950) from studies on red cells
from four healthy subjects, to be $1·718$ s.d. $\pm0·138$.

Results are affected by variations in temperature, orientation of the
red cells and, possibly, changes in the shape factor between one sample
and another. Measurements have usually been made using direct
current. If a high frequency alternating current is used the dielectric
constant of the red-cell membrane and the resistance of the red cell
adds further variability. Dellimore (1970), Ponder and Saslow (1930)
and Sirs (1968) found that the total cell volume of red cells measured
by this method was about 10% greater than when measured by any
method employing centrifugation.

The second method which is widely used is based on the interference
with conductivity as red cells pass through a small orifice (100 μm).
This is the method used in Coulter and similar counters. In this method
it is necessary to calibrate the apparatus. The usual way is by means of a
reference preparation of partially fixed human red cells, the predetermined MCV value of which is based on microhaematocrit measurement. Subsequently, measurement of the cell volume of blood samples
by haematocrit appears to be in proportion to the Coulter values
provided that the proportion of trapped plasma is the same as that
which occurs in the original sample used for calibration. It is apparent
that this method of calibration is invalid when plasma trapping in the
test sample differs from that in the reference preparation, as for example
in iron deficiency anaemia and sickle-cell disease (see chapter 8). It is,
however, not clear whether variation in trapped plasma is the only
cause of discrepancies between the two methods. Unfortunately, the
relation between pulse amplitude and cell size is not absolute and from
an exhaustive experimental study, Helleman (1972) concluded that it
is not possible to predict the cell volume of a particle from physical
measurements of the apparatus without using a primary reference.
The accurate determination of PCV and related parameters of red cell
volume remains a problem.

V. Conclusions

A reference preparation is required with the same characteristics
but greater stability than fresh blood. It should have defined values
concerning the red-cell volume occupied after packing and as determined by dilution techniques. It would be premature at present to
define its volume independently by measurement of electrical conduc-
tivities.

A material to be used as reference for conductivity methods of measuring cell volume poses special problems. At present the practice is to define the cell volume by reference to fresh human blood centrifuged for a stated time and force. When possible dilution techniques should be used to define the cell volume of both the reference material and the fresh blood against which the reference material has been compared. A promising method is to use blood samples from a selected normal population, as mean MCV and MCHC of such a group varies only slightly in health over long periods of time.

Because centrifugation involves plasma trapping to a greater extent in some diseases than others, cell volume so measured should be clearly stated. The terms *packed* cell volume and *haematocrit* should only be used for centrifugal measurements and MCV and MCHC derived from PCV might have a prefix such as PMCV or PMCHC.

Similarly, where conductivity is used to measure cell volume, the reference material, against which the comparisons are made, should be given.

REFERENCES

Chaplin, H. and Mollison, P. L. (1952). *Blood* **7**, 1227–1238.
Dacie, J. V. and Lewis, S. M. (1968). "Practical Haematology", 4th edn. J. and A. Churchill Ltd, London.
Dellimore, J. W. (1970). Ph.D. thesis. University of London, England.
Garby, L. and Vuille, J. C. (1961). *Scand. J. clin. lab. Invest.* **13**, 642–645.
Helleman, P. W. (1972). M.D. Thesis. University of Utrecht.
Hirsch, F. G., Texter, E. C., Wood, L. A., Ballard, W. C., Moran, F. E. and Wright, I. S. (1950). *Blood* **5**, 1017–1035.
Ponder, E. P. and Saslow, G. (1930). *J. Physiol.* **70**, 18–37.
Sirs, J. A. (1968). *Biorheology* **5**, 1–14.
Stewart, G. N. (1899). *J. Physiol.* **24**, 356–373.
Stewart, J. W. S. (1966). *Bibl. haemat.* **24**, 101–106.
Velick, S. and Gorin, M. (1940). *J. gen. Physiol.* **23**, 753–771.

8. Critical Appraisal of the PCV

I. CHANARIN

Northwick Park Hospital and Medical Research Council Clinical Research Centre, Harrow, England

I. Introduction

Estimation of the packed cell volume (PCV)* of a blood sample is a highly reproducible procedure. This has led to the belief that it is also a reliable and sensitive measure of the red-cell state. There have been few reasons to challenge this assumption. Nevertheless those who have been interested in estimation of blood volume soon came to realize that red-cell mass could not be deduced from plasma volume and PCV values. The PCV of a venous sample appeared to be different from "whole body" PCV.

More recently, values for the mean corpuscular haemoglobin concentration (MCHC) recorded with automatic blood counting equipment in hypochromic anaemia were found to differ significantly from those obtained with manual methods. Since haemoglobin values by the two methods were identical the differences could only be ascribed to variation in PCV results. Thus it is appropriate that a critical re-appraisal be made of the PCV.

* Or packed red-cell volume, PRCV (see p. 98).

II. Interpretation of Variance of the PCV

The PCV is one of the basic values required in calculating absolute indices relating to red-cell size (mean corpuscular volume, MCV) and concentration of haemoglobin in red cells (mean corpuscular haemoglobin concentration, MCHC). The general availability of automatic blood counting equipment has made it possible to obtain accurate red-cell counts and highlighted the value of absolute indices in haematological diagnosis. The difficulty in diagnosis of early macrocytosis or early hypochromia from inspection of blood films is generally appreciated. These difficulties have been largely resolved by automatic counting data and the mean corpuscular volume when interpreted in conjunction with the red-cell count has emerged as an exceptionally valuable diagnostic tool.

(1) An elevated MCV is the earliest evidence of a megaloblastic process. An elevated MCV is also present in over 50% of patients with myxoedema and in over 80% of chronic alcoholics presenting to their doctors, (Wu *et al.*, 1974). Macrocytosis is present in marrow failure, sideroblastic anaemia, in drug-induced red-cell dyserythropoiesis and in chronic haemolytic states.

(2) A reduced MCV (and MCH) is found in (i) iron-deficiency anaemia; (ii) anaemia of chronic disorders; (iii) α and β thalassaemia trait. Thalassaemia is likely if the red-cell count is greater than $5 \cdot 5 \times 10^{12}$ l and iron-deficiency if it is less than $5 \cdot 0 \times 10^{12}/l$.

(3) Red-cell size is of particular value in pregnancy where a variable fall in haemoglobin concentration due to haemodilution may be difficult to differentiate from iron-deficiency anaemia. Normal MCV, irrespective of the haemoglobin level indicates haemodilution; a low MCV the disorders listed under (2). A slowly rising MCV in pregnancy indicates folate deficiency and the early onset of a megaloblastic process.

(4) In malnourished populations or in populations whose dietary habits are strictly vegetarian, such as Hindus, or in hookworm infested areas, the MCV is invaluable both in diagnosis and as a guide as to whether to proceed to more elaborate investigations such as microbiological assay for vitamin B_{12} or folate.

Since the MCV is derived from PCV/red-cell count the precise values depend on the accuracy of these measurements. Automatic red-cell counting is now both accurate and reproducible in competent hands and variation in MCV results is dependent on derivation of a PCV value. This remains so despite the fact that in electronic counters of the Coulter Model S type the MCV is assessed directly and independently of the PCV. The design of the machine is such that values are adjusted so that the MCV conforms to the value obtained if red-cell count were divided into PCV. In other machines such as the Hemac 630L (Ortho), the PCV is derived directly from red-cell size and the MCV computed.

The second "absolute" value, the mean corpuscular haemoglobin (MCH), largely depends on red-cell size increasing in value in macrocytosis and decreasing in microcytosis. In severe iron-deficiency anaemia the haemoglobin content of the red cell declines at a greater rate than the fall in cell size so that there will be a greater fall in MCH than in MCV and under these circumstances a fall in the MCHC.

The third parameter, the MCHC, like the MCV, is also dependent on the PCV for its derivation. When the PCV is determined by centrifugation a fall in the MCHC is a relatively early indication of failure of haemoglobinization of red cells. With an electronically derived PCV however, the early fall in MCHC is not seen. This is because a centrifugation PCV always included a variable amount of plasma trapped between the red cells, whereas an electronically derived PCV does not necessarily do so. These differences in MCHC values have highlighted the basic problem in the use of the PCV in the laboratory.

B. PCV AND THE COLLECTION OF THE BLOOD SAMPLE

The PCV can vary with the manner and timing of the blood collection. These changes are due to relatively rapid changes in plasma volume induced by variation in posture, exercise, bed rest, meals, as well as local stasis.

Posture. The PCV is less in ambulant subjects and higher after a period of recumbancy (Walters, 1933). The difference in the same subject can vary from 2–6 divisions in the PCV value (about 5–14% of the red-cell column). The variations are greater in oedematous subjects (Fawcett and Wynn, 1960).

Stasis. A tourniquet left on the arm for more than 1 min increased the PCV by 0·4 divisions, after 2 min by one division and when left for 3 min, about three divisions (Berry et al., 1950).

Exercise. Vigorous exercise for 30 min increased the PCV between 2–5 divisions (5–12%), (Beaumont et al., 1972).

Bed rest. There is a significant rise in PCV after 2 weeks in bed, the average rise being 1·6 divisions (3·7%), Beaumont et al., 1972.

Food. An 800 calorie meal is followed by a fall in PCV apparent after about 30 min, maximal at 2–4 h and disappearing after 6 h. The fall is of the order of 1·5 divisions (Mayer, 1965).

All these factors act by changes in plasma volume, plasma protein and/or changes in osmolarity. The change in PCV is accompanied by corresponding changes in red-cell count and haemoglobin concentration and hence there is no effect on the absolute values.

Anticoagulant. Excess of anticoagulant causes shrinkage of red cells and a lowering of the PCV. This situation may arise when too little blood is added to a container designed to take a certain proportion of blood to EDTA. Heparin does not have this effect.

C. PCV AND VARIATION IN TECHNIQUE

As discussed in the previous chapter (p. 98) when blood is centrifuged in a narrow bore tube the PCV reading depends on the centrifugal forces applied provided sufficient time is allowed for centrifugation. The older type of Wintrobe tube has been largely supplanted by capillary type glass using a specially designed centrifuge which provides a centrifugal speed of 11 000 rev./min. With a radius of 9 cm from the centre of the centrifuge to the end of the capillary, a force of 12 000 *g* is applied. With a normal PCV maximum packing of red cells is achieved in 5 min and with an elevated PCV within 10 min.

Variation in centrifugation conditions can produce considerable changes in the PCV not only due to variation in the amount of plasma trapped in the red-cell column but also to variation in packing of the red cells. Sirs (1968) has demonstrated the extent to which compression of the red cells may cause error. Thus, for example a microhaematocrit PCV value of 0·455 corresponds to a true PCV of 0·525. It was also suggested that variation in flexibility of red cells influence PCV values. Defibrination of blood with glass beads left red cells relatively inflexible as compared to blood collected in anticoagulant (Rampling and Sirs, 1970).

PCV determined by a microhaematocrit method is marginally lower using capillary blood as compared to venous blood. The difference is of the order of 0·5–1·5 divisions (McGovern *et al.*, 1955). The Wintrobe method with venous blood gives a PCV which is about two divisions higher than that with microhaematocrit methods. Errors may also arise from difficulty in reading the results such as defining the two ends of the red-cell column and in the case of Wintrobe tubes, by the use of tubes with sloping ends.

III. PCV AND PLASMA TRAPPING

Mention has already been made of the fact that a variable amount of the red-cell column in the PCV is due to plasma trapped between the red cells. Two methods have been used generally in measuring this trapped plasma. The first is the addition of a trace amount of dye

(T-1824) which adsorbs to plasma albumin. The amount of dye in the plasma column is compared with that in the red-cell column (Gregersen et al., 1935; Chaplin and Mollison, 1952). The second is the addition of trace amounts of albumin labelled with radioactive iodine (Leeson and Reeve, 1951; Vazquez et al., 1952).

The higher the PCV the more plasma is trapped (Chaplin and Mollison, 1952). Further, more plasma is trapped in the upper layers of the red-cell column than the lower layers (Leeson and Reeve, 1951).

With microhaematocrit methods and normal bloods the average amount of trapped plasma expressed as a percentage of the total red-cell column has been found to be 1·3% (Garby and Vuille, 1961), 2·78% (Rustad, 1964) and 3·2% (England et al., 1972). The higher values correspond to 1·3–1·4 divisions in the PCV.

Far less attention has been paid to plasma trapping with abnormal bloods. Chaplin and Mollison (1952) noted less plasma trapping with macrocytic red cells than microcytic ones, although their figures do not show any significant differences between abnormal and normal bloods. Furth (1956) found more plasma trapping with blood in hereditary spherocytosis as compared to normal blood. England et al. (1972) compared plasma trapping in various disorders (Table I).

TABLE I. Trapped plasma as percentage
of red-cell column

No. of subjects	Range	Mean
Controls (26)	0·8–4·2	3·2
Hypochromia (14)	3·6–6·2	4·4
Macrocytosis (14)	2·4–5·7	3·7
Sickle-trait (1)	5·2	—
(fully sickled)	22·4	—

There was a highly significant correlation between MCH (and presumably MCV) and plasma trapping in the hypochromic group.

The effect of correction for trapped plasma on the normal blood count is shown in Table II (England et al., 1972). The PCV and MCV are lower when trapped plasma is excluded and the MCHC is higher.

Of greater significance is the effect of increased plasma trapping with the increase in severity of hypochromia. The marked decline in MCHC to values as low as 22 g/dl in severe iron-deficiency anaemia is due to increasing plasma trapping and a falling MCHC becomes a much later feature of iron deficiency when the PCV is derived by

TABLE II. Normal mean haematological values adjusted to an absolute PCV

	Males (32)		Females (32)	
	Uncorrected	Corrected for trapped plasma	Uncorrected	Corrected for trapped plasma
Haemoglobin g/dl	14·7		13·5	
PCV	0·442	0·428	0·409	0·395
MCHC g/dl	33·2	33·8	33·0	34·0
MCV fl.	88·4	85·2	88·9	85·8

electronic means. The falsely high PCV also obscures the fall in MCV (but not the fall in MCH) in hypochromia.

IV. ELECTRONIC DERIVATION OF THE PCV

Various methods have been used in deriving the PCV indirectly as opposed to obtaining a centrifugation value.

Kernen *et al.* (1961) found that the electrical resistance of a column of blood is a function of the relative volume per cent of the red cells. This principle was utilized in deriving the PCV (Technicon Instruments Corporation) but experience indicated that falsely low values were obtained with anaemic bloods.

The principle used by Coulter Electronics depends on the change of electrical resistance produced when the red cell passes through a standard size aperture which has an immersed electrode on each side. This results in a voltage pulse, the magnitude of which is related to the size of the red cell. From the data the mean red-cell size (MCV) is computed. The MCV in conjunction with red-cell count is then used to compute a value for the PCV. The design of the larger machines is such that the actual value recorded is adjusted to the user's ideas on what constitutes a correct PCV. The values can correspond to values obtained with blood samples of known values. In this way the PCV corresponds to centrifugation values with trapped plasma *included*. Alternatively the machine can be set to a value that excludes trapped plasma when the values in Table II are obtained. In the latter case, the PCV closely approximates to the true value in most normal blood samples but is unable to compensate fully for increased plasma trapping of abnormal bloods, particularly iron-deficient ones.

A third system (Ortho) measures the pulse generated by the passage of the red cell through a laser beam. The magnitude of the pulse is related to the volume of the cell and these are summated to produce a PCV. There is little published experience with this system which in theory is capable of reading an absolute PCV which cannot be influenced by the problems that beset centrifugation or other systems. The MCV is computed from the PCV and red-cell count.

Although the literature abounds with publications showing the identity of results between manual and automated blood counting systems, significant differences appear which are most apparent in the MCHC, as discussed above.

V. PCV AND WHOLE BODY PCV

The value of red-cell volume/red-cell volume + plasma volume is not the same as the PCV on a venous or capillary blood sample. This difference is not accounted for by the trapped plasma included in centrifugation methods for estimation of the PCV. Even when the PCV is reduced to exclude trapped plasma the PCV value still gives too high an estimate for the proportion of red cells to plasma in the body. Blood in the larger vessels have less plasma to red cells than blood in small vessels in the tissues and organs.

The implication is that blood volume cannot be calculated by estimating either plasma volume alone or red-cell volume alone and deducing the other value from the ratio of red cells to plasma in the PCV. It has been suggested that the body haematocrit/venous haematocrit ratio is 0·91 (Chaplin et al., 1953) and that when the venous PCV is adjusted in this way whole blood volume can be deduced from knowing the plasma volume alone. Others have suggested that the body haematocrit/venous haematocrit is near 0·86 (Kirsh et al., 1971) or has a wide range of 0·85–1·00 (Strumia et al., 1968). The use of this ratio however, has been criticized as given unpredictable and often serious error in calculating red-cell mass and such errors were particularly evident when the red-cell mass was increased (Kirsh et al., 1971).

VI. CONCLUSION

Availability of a true PCV on a venous blood sample would lead to greater precision in reading of absolute red-cell indices, greater uniformity in results from one laboratory to another and in improved haematological diagnosis particularly in early megaloblastic anaemia and hypochromia.

A true PCV can be obtained by estimating and subtracting trapped plasma or probably by electronic blood counting machines which derive the PCV directly from a summation of values for red-cell size.

REFERENCES

Beaumont, W. van., Greenleaf, J. E. and Juhos, L. (1972). *J. appl. Physiol.* **33**, 55–61.

Berry, T. J., Perkins, E. and Jernstrom, P. (1950). *Am. J. clin. Path.* **20**, 765–767.

Chaplin, H. and Mollison, P. L. (1952). *Blood* **7**, 1227–1238.

Chaplin, H. Jr, Mollison, P. L. and Vetter, H. (1953). *J. clin. Invest.* **32**, 1309–1316.

England, J. M., Walford, D. M. and Waters, D. A. W. (1972). *Br. J. Haemat.* **23**, 247–256.

Fawcett, J. K. and Wynn, V. (1960). *J. clin. Path.* **13**, 304–310.

Furth, F. W. (1956). *J. Lab. clin. Med.* **48**, 421–430.

Garby, L. and Vuille, J.-C. (1961). *J. clin. Lab. Invest.* **13**, 642–645.

Gregersen, M. I., Gibson, J. J. and Stead, E. A. (1935). *Am. J. Physiol.* **113**, 54–55.

Kernen, J. A., Wurzel, H., and Okada, R. (1961). *J. Lab. clin. Med.* **57**, 635–641.

Kirsch, K. A., Johnson, R. F. and Gorton, R. J. (1971). *J. Nucl. Med.* **12**, 17–21.

Leeson, D. and Reeve, E. B. (1951). *J. Physiol., Lond.* **115**, 129–142.

Mayer, G. A. (1965). *Can. med. Ass. J.* **93**, 1006–1008.

McGovern, J. J., Jones, A. R. and Steinberg, A. G. (1955). *New Engl. J. Med.* **253**, 308–312.

Rampling, M. and Sirs, J. A. (1970). *Physics Med. Biol.* **15**, 15–21.

Rustad, H. (1964). *Scand. J. clin. Lab. Invest.* **16**, 677–679.

Sirs, J. A. (1968). *Biorheology* **5**, 1–14.

Strumia, M. M., Strumia, P. V. and Dugan, A. (1968). *Transfusion, Philad.* **8**, 197–209.

Vazquez, O. N., Newerly, K., Yalow, R. S. and Berseon, S. A. (1952). *J. Lab. clin. Med.* **39**, 595–604.

Walters, S. (1933). *Am. J. Physiol.* **105**, 96–97.

Wu, A., Chanarin, I. and Levi, A. J. (1974). *Lancet* **i**, 829–831.

9. A Statistical Approach to Quality Control

BRIAN S. BULL

Department of Pathology and Laboratory Medicine,
Loma Linda University School of Medicine, Loma Linda,
California 92354, U.S.A.

I. Introduction

Data for statistical quality control in haematology can be derived from two sources. Primary or secondary standards may be run along with patient whole blood samples, or the patient whole blood samples may themselves be used as the source of quality control data.

Even though the use of standards is by far the most popular approach, it presents numerous problems. In the field of whole blood haematological analyses, primary standards are available only for the haemoglobin determination. A known number of particles cannot be weighed and packed cell volume standards cannot be produced using volumetric devices and balances. The available standards lack long-term stability. This instability necessitates frequent shipment from the manufacturer so that overlapping batches will permit a laboratory to check the new batch against an older one; it also increases the cost of a process which is already expensive and it poses a virtually insurmountable logistic problem to intra-laboratory standardization.

The alternative, the use of patient data, has received much less attention, probably because it has more limited applicability as a general approach to quality control. It can be applied most successfully to those tests where the qanutity of data is large, the coefficient of variation small, and the distribution of results approximately Gaussian. In such situations, not only will methods based on patient data function, but they show numerous advantages. They are much less expensive to implement, both in time and material, since the data is produced as a by-product of the analysis of patient whole blood specimens. Problems of shipping and of keeping overlapping batches of whole blood standards on hand are eliminated. More importantly, the sensitivity of the quality control checks increases and they become available more frequently as the test load goes up. These last two characteristics are of considerable importance. They make such patient data based approaches ideal for the quality control of automated whole blood analysers in view of the vast test output potential of these machines.

II. The Characteristics of Red-cell Indices

Red-cell indices derived from patient population can serve as the data base of a quality control programme. The distribution of the data is symmetrical and approximately Gaussian, and there is no significant day-to-day or week-to-week variability of the indices of patients in medium- to large-sized hospitals (Dorsey, 1963; Bull and Elashoff, 1974). This presumably occurs because the large majority of patients in a general hospital have normal indices. Furthermore, a considerable number of anaemic patients, particularly when the anaemia is due to acute blood loss, will also have normal values.

Thus any apparent change in the mean red-cell indices of the patient population indicates a change in the calibration of the machine which is performing the analysis since stability of the mean indices of the entire population can be confidently expected (Fig. 1).

III. The Development of a Robust Estimator* of the Mean

A quality control programme that requires large numbers of patient samples will contribute to quality control only in retrospect. If determination of defects "after the fact" were sufficient for the adequate quality control of automated whole blood analysers, then uncomplicated

---

* "Robust estimator" is a statistical procedure which is relatively insensitive to departures from the assumptions on which it depends. It is, for example, independent of the skewness of data.

estimations of the mean would suffice. The arithmetic mean of sample sizes of 60 or greater will reliably determine the mean red-cell indices of the patient population, and thus by inference the state of calibration of the automated whole blood analyser. It is because the estimation of the arithmetic mean requires such a large number of samples that it is relatively ineffective as the basis of a quality control computation. There are very few labs with a sufficient number of samples so that data

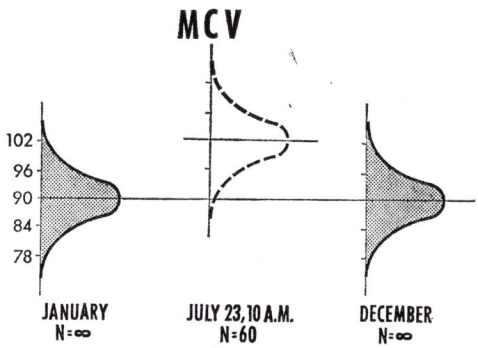

Fig. 1. The basis for the use of patient population means in quality control. A mean MCV of 102.5 on 60 randomly selected patients indicates a change in the calibration of the machine which is determining the MCV. The other alternative, that the mean of the entire patient population has changed from 90, is much less likely.

would be available more than twice during the working day. Furthermore, any defect in the machine will go unnoticed for at least as long as the time required to run a 60-sample batch and if the defect is small, it will probably go unnoticed for twice this period of time, until 120 samples have been analysed. For these reasons, if a quality control approach based on the method of patient-derived data is to be workable, a robust estimator of the mean that will function on sample sizes significantly less than 60 is needed.

An extensive analysis of several estimators of the mean indicated that when applied to data derived from red-cell indices, the best of those estimators would function adequately on a sample size of 20 (Bull *et al.*, 1974). This is a convenient number since any laboratory analysing 150 whole blood specimens every 24 h will know about the adequacy of analyser calibration at almost hourly intervals throughout the working day.

Two complementary approaches are involved in rendering an estimator of the mean more robust than the arithmetic calculation of an average value. The first, by incorporating information from preceding batches of 20 samples, "smooths" the distribution. Suppose, for example, that an arithmetic mean has been calculated on the batch illustrated in Fig. 2 (left panel). Although the mean of the patient population is 90, the arithmetic mean of this particular sample of 20 patients is higher, 90·9. Incorporation of information from the preceding batch of 20 patients (as, for example, by taking 60% of the present mean and adding to that 40% of the mean of the preceding batch which

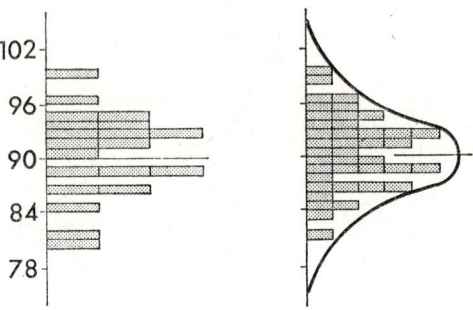

FIG. 2. "Smoothing" the data by increasing sample size. The larger the data batch the more nearly it will approximate the smooth curve descriptive of MCV data on an infinite number of patients.

in turn had incorporated 40% of its preceding batch and so on) will have a marked "smoothing" effect on the computation (Fig. 2, right panel). This incorporation of data from preceding batches is the principle underlying computations of the "moving average" or "running mean" type (Roberts, 1959).

A second approach to making an estimation of the mean more robust is to "trim" the data (Fig. 3). This can be done in several ways. The highest and the lowest values in each batch can be ignored, or any

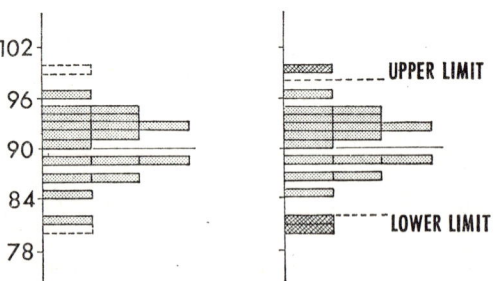

FIG. 3. Two approaches to "trimming" the data. In the left-hand panel the highest and lowest values are excluded. In the right-hand panel any values falling outside arbitrary upper and lower limits are excluded.

value falling outside of some arbitrary upper and lower limits can be excluded from subsequent calculations. However the operation is performed, the "trimming" function contributes significantly to making an estimator of the mean more robust. That is, the estimator will more reliably determine the mean of a population on small sample sizes of the order of 20 or less.

Analysis of the relative contribution of "smoothing" function and "trimming" function showed that the "smoothing" of each data batch by incorporating information from preceding data batches is quantitatively the most effective, and a simple "moving average" performed by determining the arithmetic mean of each batch of 20 samples and weighing the information from the preceding mean 40–60% of the present mean established the mean of the patient population with sufficient precision to serve as an acceptable quality control method. The addition of a "trimming" function, while not quantitatively as impressive as the "smoothing" function, did result in worthwhile improvement. The least complex and fortunately also the most effective of estimators of the mean which both "smoothed" and "trimmed" the data was the algorithm $\bar{X}_B$ (Bull et al., 1974).

Let $\bar{X}_{B(i-1)}$ denote this average after $(i-1)$ batches. Also, let X_{1i}, $X_{2i}, \ldots, X_{20i}$ denote the patient values in batch i. Then after i batches

$$\bar{X}_{B,i} = \bar{X}_{B,i-1} + \left[\frac{\sum_j \sqrt{X_{ji} - \bar{X}_{B,i-1}}}{N} \right]^2$$

when all $X_{ji} \geq \bar{X}_{B,i-1}$ and $\bar{X}_{B,0} = \mu$.

In the laboratory situation, $X_{ji} \geq \bar{X}_{B,i-1}$ does not always obtain. That is, both positive and negative values result when each incoming patient index is subtracted from the mean of the preceding batch of 20 indices. If the algorithm above is written to reflect this possibility, it takes the considerably more complex form:

$$\bar{X}_{B,i} = \bar{X}_{B,i-1} + \text{sgn}\left(\sum \frac{(X_{ji} - \bar{X}_{B,i-1})}{\sqrt{|X_{ji} - \bar{X}_{B,i-1}|}} \right) \left[\sum \frac{(X_{ji} - \bar{X}_{B,i-1})}{|X_{ji} - \bar{X}_{B,i-1}|} \right.$$
$$\left. \cdot \frac{\sqrt{|X_{ji} - \bar{X}_{B,i-1}|}}{N} \right]^2$$

This algorithm can be handled adequately on a programmable calculator provided it has at least 100 programme steps and eight storage registers. How does this algorithm "smooth" and "trim" the data? "Trimming" occurs because of the square root function. Take, for example, data on the MCV. Assume that the preceding batch mean (X_{i-1}) was 90fl. If the first patient sample in the new batch (batch i) has an MCV of 99, then the value of $X_{ji} - \bar{X}_{\text{B } i-1}$ is 9 and the square root of this is of course 3. If the MCV on the next patient sample is 91, then the square root of $X_{ji} - \bar{X}_{\text{B } i-1}$ is 1. Since these two values are now to be added arithmetically, the outlying value (99–90) will only contribute three times as much weight to the subsequent computations as did the quantitatively nine times smaller MCV (91–90). Thus the contribution of any patient MCV value to the calculation of the mean decreases the further away that value is from the mean of the preceding sample and

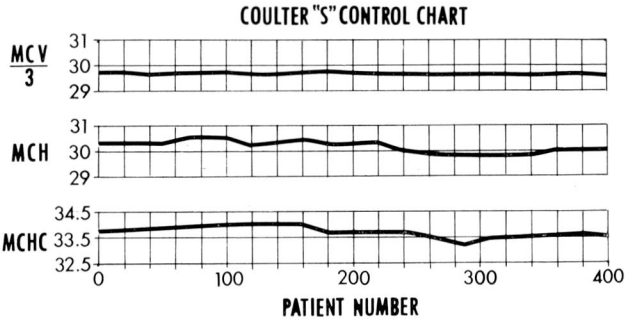

FIG. 4. A typical quality control chart plotting the mean indices determined on batches of 20 consecutive patients using $\bar{X}_{\text{B}}$.

by this means, outliers are "trimmed" progressively more severely as their deviation increases. It is because this "trimming" function takes place symmetrically around the mean of the preceding batch that the algorithm also exerts a "smoothing" function. Outliers are defined not in terms of the present batch mean but in terms of the preceding batch mean, and information from the preceding batch is carried over into the computation of the present batch mean. Thus the algorithm exerts both a "smoothing" and a "trimming" function and, when applied to the data from patient red-cell indices, produces such smooth and continuous plots as are shown in Fig. 4.

IV. The Application of the Estimator $\bar{X}_{\text{B}}$ in a Routine Laboratory

Initiation of a quality control programme utilizing $\bar{X}_{\text{B}}$ is a straightforward procedure. It will provide satisfactory quality control data both for calibration and for detection of drift on any of the large whole blood analysers such as the Coulter Counter Model S, the Technicon Hemalog, and the Hemac 630L. First the mean indices of the patient population must be determined. This is most conveniently done by taking the arithmetic mean average of several hundred patient samples. The samples should be analysed on a machine that has been carefully and rigorously standardized against commercial whole blood preparations. Alternatively a blood sample on which the red-cell and white-cell counts have been determined in triplicate using a recently calibrated digital cell counter, the PCV by means of a microhematocrit centrifuge, and the haemoglobin on a colorimeter rigorously standardized against a certified cyanmethaemoglobin standard may be used. Once the mean indices of the population have been determined on these several hundred results, then quality control charts similar to that shown in Fig. 4 are constructed and control limits are drawn 3% above and below the mean indices as determined. A desk-top programmable calculator is now programmed to accept the patient data and perform the computations required to produce new mean indices as soon as 20 patient samples have been entered. If the automated whole blood analyser is on line to a computer, then the routine can be written so that the computer accomplishes the same purposes. The graphs for each of the indices (MCV, MCH, and MCHC) are updated throughout the day as each batch of 20 patients is processed. The analyser is considered to be "in control" when mean MCV, MCH, and MCHC determined on a batch of 20 patients by means of the algorithm are within 3% of the expected mean indices of the population. All patient samples processed during the time the analyser is "in control" become, in effect, secondary standards and, if kept properly refrigerated, can be used for periods of up to 24 h to recalibrate the analyser if a breakdown causing calibration loss should occur. The secondary standards are useful under another circumstance. Occasionally a group of patient samples may all have abnormal indices such as from a cancer chemotherapy ward. Such a patient group will cause a spike in the graphs of the mean indices which will usually rapidly return to normal (see Fig. 8). At the time the patient spike occurs, it can be distinguished from machine malfunction by re-running four or five of the secondary standards referred to above. If it is a patient spike, the previous results will be duplicated; if it is machine malfunction, the results will be off by the amount of change in the machine calibration.

The implementation of this approach requires approximately 3 min for each 20-patient batch that is processed. It is possible to detect machine drift and calibration loss of as little as 1% by this approach.

V. Pattern Recognition

If all three indices are plotted on graphs such as shown in Fig. 4 throughout the day, additional information is provided by certain

patterns which occur from time to time. These patterns are characteristic for the particular whole blood analyser involved and can be
interpreted only if some basic facts about the analyser are taken into
consideration.

For illustrative purposes, four of the patterns most commonly encountered on the Model S Coulter Counter will be illustrated and
discussed. In order to interpret these patterns, it is important to
recognize that in the analysis of a whole blood sample by this machine,

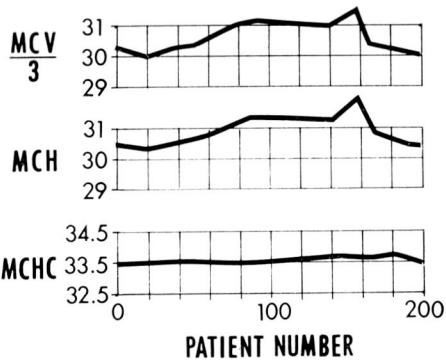

Fig. 5. The most commonly encountered pattern. Both the MCV and MCH move equally
and simultaneously in the same direction. This is caused by partial blockage and subsequent
clearing of the red-cell counting apertures.

two of the three red-cell indices are computed; the third, the MCV, is
measured directly. The red-cell count divided by the haemoglobin
value (both of which are measured on the Coulter S) gives rise to the
MCH. The PCV is derived in a more complex fashion. First, the MCV
is multiplied by the red-cell count to produce a calculated PCV and this
is divided by a measured haemoglobin value to produce the MCHC.

The patterns are all illustrated through a complete cycle of drift or
shift and subsequent recalibration. A drift has been arbitrarily illustrated for the MCV in each case.

(1) The commonest pattern is shown in Fig. 5, where both the MCV
and MCH move equally and simultaneously in the same direction.
Because of the computations involved, it is only possible for this to
occur if while the MCV is rising the red-cell count is falling. On the
Model S, both these determinations utilize the same transducer—
the red-cell counting apertures. Partial blockage of these apertures will
raise the MCV and lower the red-cell count by approximately the
same amount, thus producing congruent changes in the MCV and the
MCH.

(2) If, however, a rise (or fall) in the MCV is not the result of partial blockage of the aperture but rather due to drift in the electronic processing section, then the pattern which results consists of mirror image changes in the MCV and MCHC (Fig. 6). This occurs because

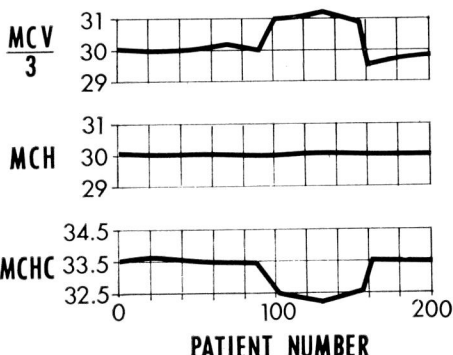

PATIENT NUMBER

FIG. 6. A mirror image pattern in the MCV and MCHC caused by electronic malfunction, usually in the MCV computation circuits.

a rise in the MCV will raise the PCV and a rise in the PCV will lower the MCHC, assuming that all other measurements stay constant. It is thus possible, by simple inspection of these patterns, not only to determine that a defect has occurred in the automated analyser but sometimes also to localize it.

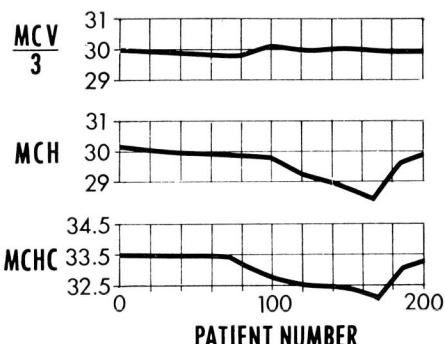

PATIENT NUMBER

FIG. 7. Simultaneous changes in the MCH and MCHC caused either by drift in the haemoglobin determination or by loss of calibration of the red-cell count.

(3) The third pattern (Fig. 7) can represent either a change in the haemoglobin determination or the opposite change in the red-cell count. The simplest way to distinguish one of these possibilities from the other is to run a blood sample on which the haemoglobin has already been

determined by reference to a standard cyanmethaemoglobin solution. Comparison of the haemoglobin value on this sample as determined by the Coulter S with its known haemoglobin content will disclose whether the haemoglobin determination or the red-cell count is at fault.

(4) Figure 8 illustrates a "spike" caused by the inclusion in a 20-sample batch of 11 macrocytic samples from a cancer chemotherapy ward. If it is feasible to exclude most such patients in a hospital by

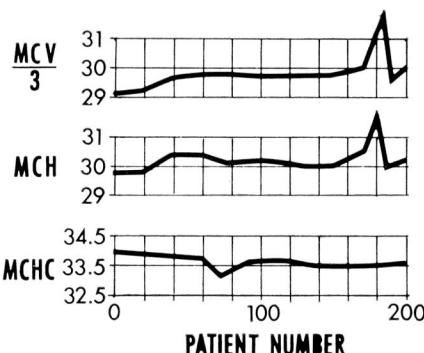

Fig. 8. A spike pattern caused by the inclusion of 11 consecutive samples from chemo-therapy patients with marked macrocytosis.

arbitrarily excluding all of the values from one or two hospital wards from the quality control computation, then this should be done. If such patients are scattered throughout the hospital, then their data is usually so diluted by data from other patients who are more nearly normal haematologically that such spike patterns will not often occur.

VI. SUMMARY

Considerable experience has shown the approximate constancy of the distribution of MCV, MCH, and MCHC values from day-to-day and week-to-week in medium-to large-sized hospitals. This supports the idea that periodic determination of the mean MCV, MCH, and MCHC of patient samples could serve as a major basis for quality control of automated whole blood analysers. The mean value of each of the indices must be determined on as small a sample as possible so that frequent checks of the adequacy of machine calibration are available throughout the day. The red-cell indices from 20 patients are sufficient if the algorithm $\bar{X}_B$ is employed. This algorithm decreases the contribution of outliers and incorporates information from the preceding group of patients. This "trims" and "smooths" the data sufficiently so that red-cell indices from 20 patients are sufficient. Graphs for each of the indices (MCV, MCH, and MCHC) are plotted and updated throughout the day as each batch of 20 patients is processed. The analyser is considered to be "in control" when mean MCV, MCH, and MCHC determined on a batch of 20 patients by use of the algorith $\bar{X}_B$ are within 3% of the expected mean indices of the population. Graphs of the behaviour of all three indices show characteristic patterns. These patterns are characteristic for each of the whole blood analysers and their interpretation depends upon a thorough understanding of the means by which each of the analysers performs its function. Given such information, the patterns can be utilized not only to determine that a defect has occurred, but also to quantitate the amount and direction of calibration loss. Sometimes when calibration loss has occurred due to machine malfunction rather than as a result of slow drift, the pattern is so highly characteristic as to pinpoint the precise location of the machine defect.

REFERENCES

Bull, B. S. and Elashoff, R. M. (1974). *Proc. San Diego Biomed. Symp.* **13,** 515–519.
Bull, B. S., Elashoff, R. M., Heilbron, D. C. and Couperus, J. (1974). *Am. J. clin. Path.* **61,** 473–481.
Dorsey, D. B. (1963). *Am. J. clin. Path.* **40,** 457–464.
Roberts, S. W. (1959). *Technometrics* **1,** 239–250.

10. Inbuilt Quality Control

J. E. PETTIT

*Department of Haematology, Royal Free Hospital School
of Medicine, London, England*

I. INTRODUCTION

The haematology laboratory has become increasingly capable of presenting the physician with highly accurate and well defined results in a shorter time than was ever before possible. Although the premise that the practising doctor treats a patient and not a series of laboratory reports still applies, it is also true that today's physician places more reliance than ever before on laboratory data. With the current emphasis on intensive treatment units the clinical laboratory does not operate with the luxury of time to ponder. Blood samples are received at a random rate throughout the day or night and the laboratory is expected to provide accurate results quickly. In many instances there is only a single chance to deliver the correct answer as therapy is initiated on the basis of the reported result.

The widely recognized forms of quality control such as the use of standards and reference preparations, the need for accurate instrument calibration and product controls, the use of statistical methods and inter-laboratory trials have gone a long way to maintain high standards of blood counting. However, using these methods alone does not constitute total quality control; no real guarantee can be given by the laboratory that a reported blood count is consistent with a particular patient at a particular time in his clinical evolution. Apart from the inherent problems of whether the day's work represented a true random distribution of values, the use of statistical methods, histograms and frequency distributions has the disadvantage that the results of the data in most laboratories are not available until the end of the day. Unfortunately, by this time it is too late to verify, delete or alter the individual blood count results. These methods are incapable of detecting the occasional "wild" errors caused by incorrect labelling, improper mixing or partial clotting of samples which are part of everyday experience in the laboratory.

Because these unpredictable "wild" errors are beyond the scope of previously mentioned quality control methods, it is essential that haematology laboratories develop a system of practice which attempts to keep the nonrecognition of such errors to a minimum. This system is "inbuilt" quality control. The method of individual result checking and the style of practice within each laboratory will reflect each laboratory director's idea of what constitutes effective "inbuilt" quality control. The author believes that most important elements in this protective operation are the use of cumulative reports, excellence in blood film examination and the clinical correlation of test results at the time of their verification.

II. The Cumulative Report

A wide variety of styles of cumulative reports are currently in use. One of the functions of these reports is to display the blood counts clearly. It is important that the design of the report allows an instant comparison of the individual parameters of the current count with those obtained on previous occasions.

A. LABORATORY MASTER CARD WITH PHOTOCOPY

The type of report which is based on a photocopy of a laboratory master card had many advantages. In Fig. 1 an example of the master card used by the Haematology Department of Hammersmith Hospital during the 1960s is shown (Dacie and Lewis, 1968). Cumulative reports similar in design to this one have proved to be valuable quality control tools. Many desirable features are incorporated. The horizontal display of sequential individual parameters which constitute the blood count allowed an instant comparison of the current result with previously recorded results. The differential leucocyte count was expressed both in percentage and absolute terms. If previous haematological data from the patient was available, each new blood count was verified only after a comparison of the current results had been made with

POSTGRADUATE MEDICAL SCHOOL HAMMERSMITH HOSPITAL	CASE No. 265211	DATE OF BIRTH 3.1.05
HAEMATOLOGY BLOOD COUNT	SURNAME SMITH	SEX M
MYELOSCLEROSIS Previous P.R.V.	FIRST NAMES JOHN / CONSULTANT DR. XYZ	WARD G5/D5

CARD NO. 3) — RACE English — DISEASE CLASSIFICATION 8.2.1. — NAME SMITH John

Date	2·4·68		6·4·68		8·4·68		17·4·68		18·4·68		19·4·68		20·4·68	
Lab. No.	10823		11257		11514		12788		12823		12977		13124	
Hb. g/100ml.	6.8		6.9		7.1		6.8		12.7		11.3		11.5	
RBC x10														
Retics. %			7.0				7.8		5.0		14		12	
PCV %	22		22		23		22		39		36		36	
M.C.H.C. %	31		31.5		31		37		32.5		31.5		32.0	
Platelets x10	394		402		380		375		401		725		980	
WBC	12,000		13,000		9,000		11,000		15,000		23,000		28,000	
	%	/cu.mm.	%	/cu.mm.	%	/cu.mm.	%	/cu.mm.	%	/cu.mm.	%	/cu.mm.	%	/cu.mm.
Blasts	4	480	7	910	5	450			6	900	4	920	3	840
Promyelocytes	4	480	4	520	4	360			2	300	7	1610	5	1400
Myelocytes	16	1920	16	2080	12	1080			9	1350	10	2300	4	1120
Metamyel.	12	1440	8	1040	8	720			16	2100	7	1610	5	1400
Neutrophils	39	4680	53	6890	46	4140			57	8550	46	10,580	53	14840
Eosinophils			2	260	2	180					2	460	4	1120
Basophils											2	460	3	840
Lymphocytes	20	2400	8	1040	19	1620			10	1500	11	2530	10	2800
Monocytes	5	600	2	260	4	360			2	300	11	2530	13	3640
NRBC's /100 WBC	2		1		3				2				5	
Film morphology	Rouleaux++ Aniso++ Poikilo+++ Frag.++ Poly + Occ giant plats.		No change		No change		I.S.Q.		Double red cell pop. Normal + patient's.		Plats ↑↑ WBC; Toxic changes Occ.Döehle Body		Occ. H.J.B. +Targets Plats ↑↑↑ Neutrophils show toxic granulation	
Signature	4.		4.		4.		LDB		4		LDB		4.	
For Lab. use only														

↑ SPLENECTOMY 4 Ⓞ BLOOD

FIG. 1. Blood count record card used by the Hammersmith Hospital Haematology Department during the 1960s. The patient whose counts are illustrated had myelosclerosis.

those of previous blood counts and after an inspection of the blood film. The diagnosis and important clinical developments were written on the card so that these facts were constantly available to the trained personnel who were responsible for result verification. In the generation of ward and patient file reports transcription error was kept to a minimum by directly photocopying the laboratory master card.

B. THE COMPUTER PRINTOUT OR CRT DISPLAY

The advent of complex electronic counters has considerably increased the accuracy and precision of red-cell and platelet counting. In parallel with this development the introduction of computers into haematology laboratories has provided an opportunity to generate excellent cumulative reports. Modern automated electronic counters provide a comprehensive list of blood count parameters including RBC and platelet counts and the computed RBC indices on all tested samples. Suitable interfacing between the electronic counters and the computers and the use of special terminals or consoles to generate reticulocyte and differential counts and the report of the blood film morphology virtually eliminates transcription error.

The basic concepts of the previous report have been incorporated into the computer report shown in Fig. 2. This type of cumulative report can either be generated for the ward and patient's file by the computer lineprinter or its relevant sections can be displayed on the cathode ray tube (CRT) of the specialized terminals that are used for differential counting, blood film reporting and result verification purposes. The horizontal display of individual tests and result verification procedures with this computer operation are identical to those previously described. With the laboratory master card and photocopy report a joining of consecutive cumulative reports allowed instant horizontal comparison of any number of individual blood count results. To be able to retain this facility with computer generated reports, it is imperative that the computer software development ensures regular positions on each report for all the separate parameters which comprise the blood count. This form of display facilitates the use of the cumulative report in quality control. Obviously, it is an easier computer operation to generate vertical style reports and many laboratories have unfortunately chosen this latter method. The author believes that the advantages of horizontal display far outweigh any minor difficulties that may be encountered during the software development of the horizontal style report.

In both examples of cumulative reports shown the primary red-cell estimations are shown at the top of the column of blood count results. The reticulocyte count is placed directly below the red-cell count and this facilitates the calculation of the absolute reticulocyte level if this count is preferred. With accurate electronic red-cell counting there would be considerable merit in expressing all reticulocytes in absolute terms rather than as percentages. However, because of the lack of general acceptance for this change most laboratories continue to report the latter figure. In the computer report, to complete the red-cell

FOOTHILLS HOSPITAL · CALGARY, ALBERTA T2N 2T9   DEPARTMENT OF LABORATORIES

```
FOOTHILLS              P A T I E N T   R E P O R T          # 1 PAGE 1
UNIT 62                    17-JAN-74    15:20

663    168713  36025554878  03-JAN-74
62     JONES   THOMAS JOHN
       DR  DOE J                20-MAR-2152
```

```
                       ACC #:  101226   101376   101438   101537   101651   101726
                       DATE :  03-JAN   04-JAN   06-JAN   09-JAN   13-JAN   17-JAN
                       TIME :  09:00    10:00    10:00    09:00    11:00    08:00
```

---HAEMATOLOGY---

HEMALOG

HAEMOGLOBIN	GM/DL	4.6	4.6	5.5	6.4	6.7	7.4
PCV	%	14	14.5	17	20	21	23
RBC	MIL/UL	1.148	1.208	1.417	1.835	1.944	2.255
RETICULOCYTE COUNT	%	1.0	1.5	7.0	21	28	19
MCV	CU U	122	120	120	109	108	102
MCH	UUG	40	38	39	35	34	33
MCHC	%	32.8	32.7	32.4	32.0	31.9	32.2
PLATELET COUNT	THO/UL	90	86	125	285	397	231
WBC	THO/UL	3.100	3.000	3.800	4.000	4.900	5.300

DIFFERENTIAL

BANDS	THO/UL	0.062	0.090	0.304	0.200	0.147	0.106
NEUTROPHILS	THO/UL	1.395	1.290	1.710	2.120	2.744	3.127
EOSINOPHILS	THO/UL	0.062	0.090	0.076	0.120	0.098	0.053
BASOPHILS	THO/UL	0.031			0.040		0.053
MONOCYTES	THO/UL	0.310	0.450	0.380	0.320	0.441	0.477
LYMPHOCYTES	THO/UL	1.240	1.080	1.330	1.200	1.470	1.484

RBC MORPHOLOGY

ANISOCYTOSIS	+++	+++	+++	++	++	++
MACROCYTOSIS	++++	++++	+++	+++	++	++
POIKILOCYTOSIS	+++	+++	+++	++	++	++
POLYCHROMASIA			+	+++	+++	++
PLATELETS ON FILM	DECR	DECR	NORM	NORM	INCR	NORM
	A					B

A. HYPERSEGMENTED NEUTROPHILS, OCC. MEGALOBLASTS: FINDINGS CONSISTENT
    WITH MEGALOBLASTIC ANAEMIA

B. GOOD HAEMATOLOGICAL RESPONSE TO HYDROXYCOBALAMIN

KEEP

FIG. 2. Computer printout cumulative report of blood counts as used by the Foothill's Hospital Division of Haematology. The blood counts of a patient with pernicious anaemia responding to hydroxycobalamin are shown.

parameters the indices are next to be shown. The reason for placing the platelet count before the leucocyte count is to be able to document the differential count beneath the total leucocyte value. There are great advantages of reporting the differential in absolute terms. This form of reporting prevents the many interpretational errors which occur when the differential percentage is read in isolation from the total leucocyte count, e.g. a high lymphocyte percentage may be erroneously interpreted as a lymphocytosis when in reality the patient has an absolute neutropenia. Absolute values are also of crucial importance to those physicians using cytotoxic drugs. The early detection of chronic lymphocytic leukemia is made easier by absolute reporting; this diagnosis should be suspected in adult patients with a persistent lymphocytosis of more than $5.0 \times 10^9/l$.

With repetitive blood counts any gross change in individual red-cell values may indicate an error. The long red-cell survival and limited capacity of the bone marrow for regeneration renders unlikely sudden changes in Hb, PCV or RBC levels during the relatively stable periods that occur with most illnesses. The patient whose chart is illustrated in Fig. 2 had pernicious anaemia and was being treated with hydroxycobalamin. The gradual and predictable increase in primary red-cell parameters, the reticulocyte peak and gradual fall in MCV are all appropriate and indicate good quality control. To illustrate further the use of the cumulative report in quality control, this patient's blood counts are continued in Table I. The first three columns of results are

TABLE I. The cumulative report in quality control

| | | 17 Jan | 21 Jan | 28 Jan | 05 Feb Not verified | |
					A	B
Hb	g/dl	7·4	8·0	9·3	14·9	7·9
PCV		0·23	0·25	0·30	0·46	0·26
RBC	$10^{12}/l$	2·255	2·551	3·333	5·111	3·291
Retics	%	19	5·0	3·0	1·0	2·0
MCV	fl	102	98	90	90	79
MCH	pg	33	31	28	29	24
MCHC	g/dl	32·2	32	31	32·3	30·3
Platelets	$10^9/l$	231	225	286	253	410
WBC	$10^9/l$	5·300	5·800	6·000	8·300	9·010

verified blood counts. The two right hand columns A and B are hypothetical blood counts which may have been obtained 1 week after the last verified result. Consider column A. A jump in Hb value from 9·3 to 14·9 g per dl in 1 week is beyond the regeneration capability of bone marrow. The possibilities to be considered must include: (1) sample error—e.g. the blood was actually taken from a different patient or a partial loss of blood from an unmixed sample had occurred; (2) operational error—e.g. incorrect machine calibration, inadequate mixing of the sample before counting, inadequate stirring of the sample due to its incorrect placement on a machine pickup area; (3) a recent blood transfusion; (4) severe dehydration. Obviously blood count A should not have been verified. If a repeat blood count on the same sample gave acceptable but different results an operational error at the time of the original count would have been suspected. If a similar aberrant result was obtained, a phone call to the patient's physician should have been initiated to have excluded possibilities 3 and 4. The final "inbuilt" quality control procedure in this example would have been a repeat sample from the patient. If more acceptable results were obtained, this would have confirmed that a sample error was the presumable cause of the original result.

With the second example shown in Table I (blood count B on the 05-FEB) the haemoglobin had dropped from 9·3 to 7·9 g per dl during the period of 1 week. This occurrence would be unusual during a full therapeutic response to hydroxycobalamin. As an explanation for this change the patient may have had a haemorrhage or, perhaps more unlikely, a haemolytic episode during that week. However, scrutiny of the cumulative report reveals another possible reason for the reversal in the previously upward trend of haemoglobin levels. The MCH values recorded show a sequential fall: 33, 31, 28, 24. It is probable that an exhaustion of initially depleted marrow iron stores had occurred leading to the development of iron deficiency which prevented a complete therapeutic response to hydroxycobalamin. If a population of hypochromic red cells had been seen during blood film examination, it would not have been unreasonable to have verified blood count B on this day.

Other parameters of the blood count should be compared with previous results in a similar fashion to that described in the above examples and verification should proceed only if the blood film appearances are consistent with the printed figures and the results correlate with the patient's clinical condition.

As well as being able to quality control the results from an individual patient the use of cumulative reports at the verification stage of blood counting may also contribute to the overall quality control of blood counting. If blood film examination and verification procedures are timed to take place soon after the actual blood counting, the methods described are capable of detecting systematic shift errors at a relatively early stage during the day's operation. Many patients in hospital have almost daily blood counts; patients in intensive care, renal and leukemia units often fall into this category. The first indication of counting inaccuracy may be a persistent similar change in figures in a batch of such patients having repeated blood counts. Personnel responsible for blood film work may draw attention to these errors at a stage before their detection would be possible by the use of standards and statistical quality control methods.

C. THE TIME RELATED LOGARITHMIC CHART

In patients with complex haematological problems the cumulative report of leucocyte values and platelet counts can be plotted on logarithmic charts. The value of such charts for illustrative and educational purposes has been recognized for a long time. A steadily increasing number of haematologists are finding that there are many advantages in utilizing these charts during their working routine. The chart shown in Fig. 3 is an example of the type of cumulative report developed by Dr D. A. G. Galton (1960) which is currently in use in the MRC Leukemia Unit at the Royal Postgraduate Medical School. A visual summary of the progression of the haematological disorder and state of the patient at any particular time is instantly available. This type of cumulative report is of value in the titration of cytotoxic therapy and may assist the physician to anticipate severe myelodepression. In Fig. 3 it can be seen that the rate of fall in leucocyte count was proportional to the amount of busulfan the patient was taking at the time.

Over the past decade these charts have also proved to be of value to people concerned with quality control in routine haematology. Aberrant results detected by comparison with standard cumulative reports are even more obvious when they are plotted in graphic fashion. In patients receiving cytotoxic therapy the changes in leucocyte and platelet counts tend to follow linear patterns when plotted on a logarithmic scale. Physicians and haematologists who use these charts as working charts may tentatively plot the "urgent" or "stat" results, which they often receive in unverified form from the laboratory, as soon as these become available. Any plot showing gross divergence from the linear trends seen during cell proliferation, cytotoxic cell killing or stable periods should immediately arouse suspicion. A leukemia unit using these charts in this fashion may be the first to inform the diagnostic laboratory of significant counting errors.

III. THE BLOOD FILM EXAMINATION

Blood film examination is considered by most haematologists to be the most important screening test for blood disease and the cornerstone of "inbuilt" quality control. Ideally film microscopy should be performed on every sample of blood submitted for counting.

Many significant and important blood disorders are not associated with alteration in haemoglobin, PCV, total leucocyte and platelet counts e.g. early megaloblastic anaemias, compensated haemolytic

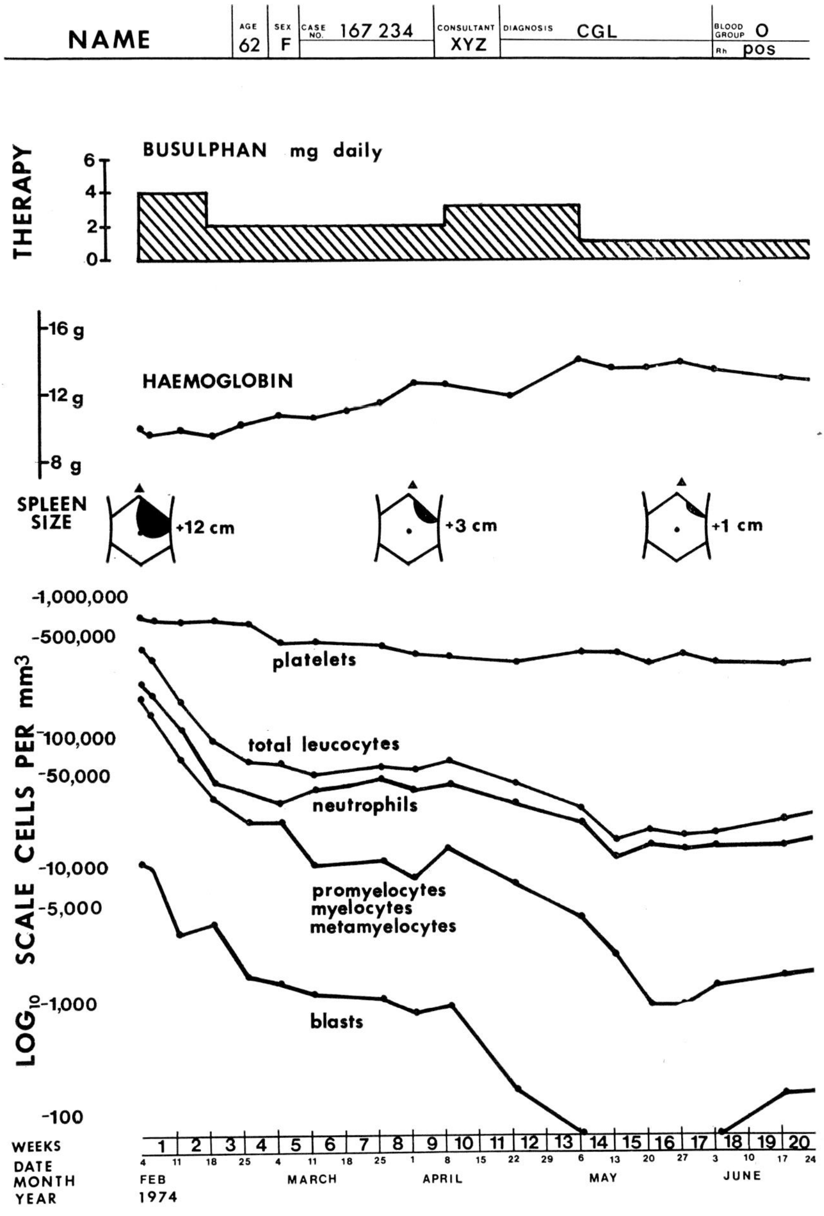

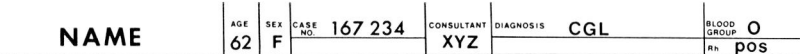

FIG. 3. Haematological chart in current use at the M.R.C. Leukemia Unit, Royal Post-graduate Medical School. The course of a patient with chronic granulocytic leukemia, following initial treatment with busulphan is illustrated.

anaemias, lead poisoning, early lymphocytic and blast-cell leukemias, infectious mononucleosis, agranulocytosis, and the eosinophilia of allergic disorders. No definite pattern of red-cell indices is found in many haematological disorders with anisocytosis and poikilocytosis, with acanthocytes and red-cell fragments or with the double population of red cells classically found in treated iron deficiency, some sideroblastic anaemias, and in the combined iron and folate deficiencies of the malabsorption syndromes; examination of the blood film is obligatory in pin pointing these blood abnormalities. The discovery of malarial parasites, filaria, and interesting hereditary abnormalities of red cells and leucocytes are additional benefits of film microscopy. The presence of marked rouleaux formation may suggest multiple myeloma or another protein abnormality and marked autoagglutination of red cells is an indication of immune haemolytic anaemia.

Blood film examination also plays an essential role in quality controlling the blood count. As such it should be the last step in routine blood counting and, as already mentioned, part of the adopted verification procedure. It is imperative to have all the counting parameters and the cumulative report available to the haematologist or technologist examining the film. Quality control checks can be made e.g. hypochromia, macrocytosis, microcytosis as indicated by the indices should be verified, polychromasia should be sought as confirmation of high reticulocyte counts, and qualitative checks of abnormally high or low leucocyte and platelet counts are mandatory. In many laboratories, especially those handling work referred in from family medical practitioners, a large proportion of the day's work involves single specimens for screening, so that reference to previous results in the form of a cumulative report is not possible. With this type of request the examination of the blood film is the only "inbuilt" quality control mechanism possible.

Because of its inherent importance in routine haematology meticulous attention should be paid to all blood film work. The blood film should be well spread, well stained and, ideally, it should be coverslipped. Examination of the film should not be a haphazard and rushed procedure. The negligent technique of dumping a drop of oil somewhere on the film and having a brief look with a 3·7 or 2 mm oil immersion objective lens should be firmly discouraged. A methodical routine of blood film examination should be demanded from the outset of haematology training programmes. Initially each film should be scanned with the 16 mm objective lens. Cells in the optimum thickness area of the film behind the tail should then be scrutinized, preferably with the 4 mm objective lens. Some laboratories prefer the 3·7 mm oil immersion lens as the standard lens for individual cell morphology and differential counts. Advocates of the high dry 4 mm technique believe that this oil immersion lens is somewhat less satisfactory in the assessment of red-cell morphology. The 2 mm oil immersion lens should be reserved for the occasional checking of fine detail in individual cells e.g. the confirmation of the nature of red-cell inclusions or stippling, of Aüer rods and leucocyte inclusions or for making a more detailed observation of leucocyte and platelet granulation or nuclear chromatin details.

Coverslipped blood films become an excellent permanent record. Blood film storage allows retrospective inspection if interesting subsequent haematological developments occur in a patient. Not only does storage allow the evolution of haematological disorders to be studied but also missed findings may be uncovered allowing continual constructive criticism within the laboratory. Although the latter is often a humbling experience it tends to maintain an edge of excellence in the blood film section; laboratory personnel tend not to take dangerous time saving short cuts during blood film examination if they are aware that the significant changes missed by such negligent techniques will always be available for others to see.

During blood film examination the reason for abnormal leucocyte and platelet counts may be found to be artefactual. Accurate differential counting is impossible to perform on unevenly spread blood films. In thin films with irregular tails the majority of neutrophils are often carried to the tail of the film with the spreader, leaving a preponderance of small lymphocytes in the part of the film where the differential would normally be counted. Differential counts done on such films often result in spurious neutropenia. It should be emphasized, though, that with stricter attention to the quality control of blood film spreading this error would be avoided. Artefactural neutropenia is sometimes the result of partial clotting of the sample and blood film examination may also reveal this cause.

Three important causes of pseudothrombocytopenia are readily detectable during blood film examination. The most common cause of spurious thrombocytopenia is partial clotting of the sample. In such cases an inspection of the tail of the blood film with the 16 mm objective lens usually reveals small clots which are composed of fibrin strands and countless numbers of aggregated platelets. Appearances such as these should initiate a repeat sample from the patient and the current blood count associated with this clotted sample should be discarded. When blood film examination fails to show the presence of small clots in blood counts with unsuspected thrombocytopenia the blood sample should be probed for larger clots. If such a larger clot is detected, again the blood count results must be discarded. In occasional patients with spuriously low or fluctuating electronic platelet counts aggregates of platelets are seen during blood film examination. Watkins and Shulman (1970) attributed this type of pseudothrombocytopenia to platelet cold agglutinins active at room temperature. More recently Shreiner and Bell (1973) reported six patients exibiting this phenomenon, but unlike Watkins and Shulman's patients the agglutinins were active up to 37°C and were dependent upon the blood being taken into the anti-coagulant ethylenediaminetetra-acetic acid (EDTA) for their activity. Mant *et al.* (1972) described two further patients with apparent thrombocytopenia and agranular platelets due to EDTA dependent platelet aggregation. Another widely observed, but infrequently reported, form of pseudothrombocytopenia results from platelet neutrophil adhesion. The cause of this *in vitro* adhesion is unknown. In the blood films or wet preparations made from blood taken into EDTA rosettes of platelets can be seen around the neutrophils. This phenomenon is not observed in films made from native blood or from blood taken into other anticoagulants. From personal observations of patients showing this change and from reviewing the cases described in the rather sparse literature on this subject (Field and MacLeod, 1963; Signy and Green, 1963; Bolton and Boyd, 1963; Crome and Barkhan, 1963; Prchal and Blakely, 1973) it is apparent that neutrophil platelet adhesion occurs both in normal individuals and in patients with widely differing medical conditions. If the cause of low electronic platelet counts has been identified during blood film examination as being EDTA-dependent platelet agglutination or neutrophil-platelet adhesion, this event should be separately documented on the report. For clinical purposes the low platelet count may be discounted and the result substituted with a manual platelet count initiated from native blood diluted in formal-citrate or ammonium oxalate.

If important haematological changes have been noted in the blood count and film examination, the final responsibility of the laboratory is a brief interpretation of these changes for the benefit of the patient's doctor. If possible a diagnosis should be inserted into the report (see comments A and B in Fig. 2).

IV. CLINICAL CORRELATION

The final checking procedure available in "inbuilt" quality control is clinical correlation. Wherever possible all significant haematological changes should be checked against the patient's history, physical findings and important clinical developments before these changes leave the laboratory as a verified report.

In many laboratories blood-count request forms have been designed with spaces for relevant clinical data and reference to such information may allow verification of results which otherwise seemed improbable. However, because of the many demands made on the time of practising doctors it is indeed a fortunate laboratory where this data is regularly furnished with the request forms.

The telephone should be considered an integral part of the hardware necessary at the blood film bench and should be used to obtain relevant clinical information if there is doubt about the validity of blood-count results. Ward consultation may explain results which remained unlikely after reviewing the cumulative report and blood film, e.g. recent haemorrhage or blood transfusion may explain dramatic changes in haemoglobin levels. One may discover that the sample had been taken out of an arterio-venous shunt or from an arm into which intravenous infusion had been running. The commencement of radiotherapy, cytotoxic drugs, or drugs known to be associated with blood dyscrasias are important facts that should be available during the final checking procedure.

An essential part of quality control in haematology is to define the significance of blood changes. Discussion between the haematologist and the patient's doctor often proves to be mutually beneficial. Clinical information provided by the patient's doctor may allow the haematologist to arrive at a diagnosis which would not have been possible with the blood-count results alone. The patient's doctor receives the benefit of the haematologist's opinion as to the significance of the blood changes or a definitive diagnosis at an early stage and the decisions as to which further investigations are appropriate may be made during this important telephone call.

If doubt exists about the validity of counts after clinical correlation the laboratory may opt to repeat the count on the original or a freshly-drawn sample before a decision is made to verify or delete and replace the disputed result.

REFERENCES

Bolton, F. G. and Boyd, J. (1963). *Br. med. J.* **ii,** 747.
Crome, P. E. and Barkhan, P. (1963). *Br. med. J.* **ii,** 871.
Dacie, J. V. and Lewis, S. M. (1968). "*Practical Haematology*", 4th edn, p. 160. Churchill, London.
Field, E. J. and MacLeod, I. (1963). *Br. med. J.* **ii,** 388–389.
Galton, D. A. G. (1960). In "*Proceedings of the VIII International Congress of Hematology*" p. 467. Pan Pacific Press, Tokyo.
Mant, M. J., Doery, J. C. G. and Gauldie, J. (1972). *Clin. Res.* **20,** 934.
Prchal, J. T. and Blakely, J. (1973). *New Engl. J. Med.* **289,** 1146.
Shreiner, D. P. and Bell, W. R. (1973). *Blood* **42,** 541–549.
Signy, A. G. and Green, A. E. (1963). *Br. med. J.* **ii,** 624.
Watkins, S. P. and Shulman, N. R. (1970). *Blood* **36,** 153–158.

11. Quality Control of Qualitative Test in Haematology

D. W. PENNER and C. C. MERRY

Departments of Pathology, University of Manitoba and
Health Sciences Centre, Winnipeg, Canada

I. INTRODUCTION

Qualitative procedures in haematology, in contrast to quantitative procedures, possess a number of unique attributes. The latter are basically numerical, chemical or physical determinations while the former are predominantly subjective value judgements, generally made on the morphologic attributes of cells, which lead to a definitive diagnosis. These interpretations (diagnoses) may, and often do, have a very direct and significant effect on patient care. While the quantitative procedures also affect patient care, they are usually only part of the total data which lead to definitive diagnosis and treatment.

Quality control programmes including the required quality evaluations (proficiency testing) have long been an integral part of haematology laboratory operations. While not perfect, these are substantial in relation to many quantitative procedures. By contrast, quality control programmes including proficiency testing are only now being developed for the qualitative procedures. All quality control programmes require reference standards and our inability to develop working value judgement standards has been the major deterrent to the development of acceptable programmes. It is obvious that different approaches to quality control of qualitative tests must be evolved in order to provide the needed assurance that this aspect of laboratory function is adequate for good patient care.

SOME ASPECTS OF QUALITATIVE PROCEDURES IN HAEMATOLOGY

It is important to define and fully appreciate the nature of the qualitative test. All qualitative procedures are based on an expression of judgement of the observer and, as such, are based upon human faculties. Unfortunately for the purposes of science, human faculties remain incapable of precise evaluation, particularly by the use of statistical methods. It is an old adage that medicine is both an art and a science and the control of the quality of qualitative tests is by appreciating the art of medicine and attempting to improve upon it. Many tests have both qualitative and quantitative aspects. Clearly, in haematology qualitative procedures are largely concerned with examination of peripheral blood films and bone marrow aspirates. The qualitative aspects are less easily recognizable in other laboratory tests. The estimation of fibrin degradation products is expressed in micrograms per millilitre and it looks very scientific and quantitative. Yet because it is a microhaemagglutination inhibition technique and accuracy is ultimately determined by the ability to recognize the end point of agglutination correctly, the results are dependent upon human judgement. The estimation of haemoglobin would seem to epitomize the quantitative test: the molecular weight has been determined, the structural formula has been established, and the millimolar extinction coefficient carefully measured. Even then human judgement must on occasion intervene and recognize potential errors due to the turbidity of hyperlipemia and nuclear debris. While we cannot measure value judgements precisely a great deal is known about the various factors which affect their reliability. The ultimate "correct" value judgements relating to biologic phenomena must predict the sequence of events and the final outcome. In the field of medicine this means the ability to recognize disease, or its absence, and the sequence of events as the disease progresses (including effects of treatment) whether to recovery, disability or death. Value judgements can therefore only be as accurate as our knowledge and understanding of the disease process under consideration. The present state of our knowledge of disease processes therefore is a significant factor affecting the accuracy of value judgements. Other factors of paramount importance include training and experience of the observer and it is generally accepted that accuracy of value judgement can be improved by additional experience and training. In haematology as in histopathology and cytology the availability of "other pertinent clinical data" may have a very significant effect on value judgements. Also of the greatest importance is the availability and the quality of the various technical procedures to which the material may be subjected. The quality of film preparation, fixation and staining, the use of special stains and techniques such as the peroxidase reaction, periodic acid Schiff reaction and Sudan black stains can greatly enhance the reliability of value judgements.

Finally, yet another aspect of value judgement relates to the "human factors", Fatigue, psychological pressures, illness, monotony, interruptions and distractions, time available to perform the task, motivation etc. may at times significantly affect the accuracy. When quantitative laboratory procedures are involved today's machine-oriented technology may minimize the effect of these human factors but they remain important in making value judgements. Since the primary concern relates to the quality of patient care, the significance of these "human factors" can only be evaluated if quality control programmes including proficiency testing, relate to examination of the qualitative procedures as they are actually performed on a day to day basis and as this performance affects patient care.

II. Some Aspects of Methodology of Quality Control and Quality Evaluation in Qualitative Procedures in Haematology

An adequate quality control programme must have two basic components: a continuous internal quality control component which must be an intrinsic part of the laboratory operation and an external quality evaluation to provide additionally needed assurance that a good quality performance, as it affects patient care, exists.

The internal quality control programme is basically one of an ongoing learning and checking process which includes continual internal and external consultations, follow-up on all cases diagnosed as abnormal, review of previous material when a subsequent abnormal diagnosis is made, use of known and unknown diagnostic check material, random review of material, compilation of statistical data to establish patterns of diagnoses, reference to other collateral evidence from clinical and other laboratory data.

Many of the internal quality control devices outlined above which undoubtedly have made a very significant contribution to the quality of our work are of questionable value in determining the actual quality of performance on a day to day basis and the effect on patient care.

For many years various external programmes have been available.

These consist of peripheral blood or bone-marrow films distributed on a continuing and regular basis. These are evaluated and the results compared with the "correct interpretation" which has been established by an individual expert or panel of experts. The experts' opinion may be only an opinion or it may have been substantiated by the outcome of the clinical case, thus increasing the probability that the "correct interpretation" is indeed accurate. Unquestionably the procedure provides a most valuable learning experience. Its use to assess accurately the quality of value judgements as practised on a day to day basis and their effect on patient care has yet to be determined. The examination of unknown material produces an artificial situation and unfortunately may not reflect the routine day to day operation. On the one hand assessment is limited by the absence of additional information, repeat films, special stains etc. On the other hand assessment may be improved by modification of the "human factors" under examination conditions including a more time consuming and meticulous study, reference to relevant articles or books or, in other words, introducing any variation from the routine performance.

To evaluate qualitative procedures as they are actually practised it is necessary to devise a method using material from patient files and assessing day to day performance as it has affected patient care. This is a retrospective means which may prospectively improve the performance of the qualitative laboratory procedures. Models using this method for histopathology and cytology have already been devised and tested, and field trials of proficiency testing are currently being undertaken in North America by the College of American Pathologists (Penner, 1973a, b).

Histopathologic and cytologic material from the patient files was examined by panels consisting of peers. The panel of peers was a group of ten practising pathologists who contributed material from their own hospital files and who then examined this pool of material as unknowns. The concurrence or lack of concurrence with the original diagnosis and the degree of concurrence within the group was established. Consensus is defined as more than 50% agreement of the peer group, the degree of consensus can therefore vary from 51–100%. The degree of consensus necessary for good patient care has yet to be established.

The logistics of obtaining patient file data and circulating it to a peer panel while requiring a considerable amount of time, depending on the geographic location of peers, offers no particular problems. Evaluation of the now considerable amount of data accumulated and the establishment of a significant consensus standard which is of import to the patient and which provides the needed assurance of quality performance will require more time and study as will the ultimate effect of this type of quality control on patient care. Also yet to be determined is the comparison of performance of peer panels and expert panels. This is currently being tested. Equally important is to determine the necessity of using patient file material in proficiency evaluation, in contrast to the use of diagnostic sets of slides. It might be possible to compile sets of slides which through critical evaluation by experimental trial, demonstrate their ability to evaluate performance as it is practised on a day to day basis. Attempts are currently being undertaken by the College of American Pathologists in this area.

III. A PROPOSED QUALITY CONTROL MODEL FOR QUALITATIVE PROCEDURES IN HAEMATOLOGY

A. INTERNAL QUALITY CONTROL COMPONENT

This consists basically of a continuing educational and checking programme. Most of the following should be included in such a programme.

(1) Participation in reading programmes, haematology seminars, short course and use of slide study sets.

(2) Follow-up system for all abnormal peripheral blood and marrow films with final confirmation of diagnoses by additional evidence such as histologic or chemical studies, and/or the clinical outcome of the case.

(3) Review of all previous films from a patient when abnormalities are detected.

(4) Maintenance of active consultation in problem cases with records of these consultations.

(5) Annual review of the pattern of diagnosis to detect alteration in these patterns.

(6) Review of randomly selected peripheral blood and marrow films previously judged to be normal.

(7) Routine use of collaborative data to check interpretations, e.g. plasma proteins in myeloma.

(8) Records of all the above activities should be maintained and available to any inspection agency concerned with accreditation.

B. EXTERNAL QUALITY CONTROL COMPONENT (PROFICIENCY TESTING)

This component must provide the needed assurance that the quality of performance, as it is practised on a day to day basis, and in its effects on patient care, is good. It is suggested that the standard used to measure the quality of performance be based on consensus standards.

The recommended model

(1) Groups of haematologists are organized so that they become both contributors and evaluators. A panel of ten contribute randomly selected material from their patients' files, and in turn evaluate this pooled material as unknowns. The use of a standardized nomenclature is essential. From this data may be derived (a) the concurrence with the original interpretation and group consensus; (b) the level of consensus required to establish the measure of acceptable individual performance in order to achieve the desired goal of improved patient care; (c) the number and composition of patient file material needed to evaluate performance.

(2) Panels of recognized experts should be established so that comparison between performance of peer panels and expert panels can be determined.

IV. SUMMARY

Qualitative tests in contrast to quantitative tests are based entirely on value judgements and hence quality control programmes require different approaches and techniques. The major qualitative procedure is the evaluation and interpretation (diagnosis) of peripheral blood and marrow smears. These often have a major direct significance to patient care. Consensus standards based on peer review of material from patient files is considered the best approach presently available. While yet to be fully evaluated, considerable experience in quality control of histopathology and cytology is now available. This is based on past work and continuing programmes of the College of American Pathologists, and it is suggested that a similar approach for qualitative tests in haematology should be used.

REFERENCES

Penner, D. W. (1973a). *In* "Pathology Annual", Vol. 8, pp. 1–19. New York, Appleton-Century-Croft.

Penner, D. W. (1973b). *In* "Proceedings of International Conference on Standardization of Diagnostic Material", Atlanta, Georgia, June 1973, pp. 267–271. U.S. Department of Health, Education and Welfare.

12. Purification, Standardization and Quality Control of Romanowsky Dyes

D. WITTEKIND and W. LÖHR

Anatomy Institute, Freiburg University,
Brsg. West Germany

I. INTRODUCTION

The term "Romanowsky dyes" (RD) is in common use but opinions are often vague as to exactly what it means. Mixtures of the anionic dye eosin (Fig. 1) with certain cationic thiazine dyes (Fig. 2) give the Romanowsky effect. This requires the presence of one or more of the demethylated homologues of methylene blue (Mbl) which occur when the Mbl undergoes oxidation (polychroming). The history of RD has been reviewed by Baker (1958) and by Harms (1965), and a useful "glossary" of the RD-components appeared in a paper by Cramer *et al.* (1973).

Following the original work by the Russian protozoologist Romanowsky (1891) who had demonstrated the nucleus of the malaria parasite by adding a 1% solution of eosin to a saturated solution of Mbl, and that of Nocht (1898) who concluded that polychroming of the Mbl was responsible for the staining effect, Giemsa (1902, 1904, 1907)

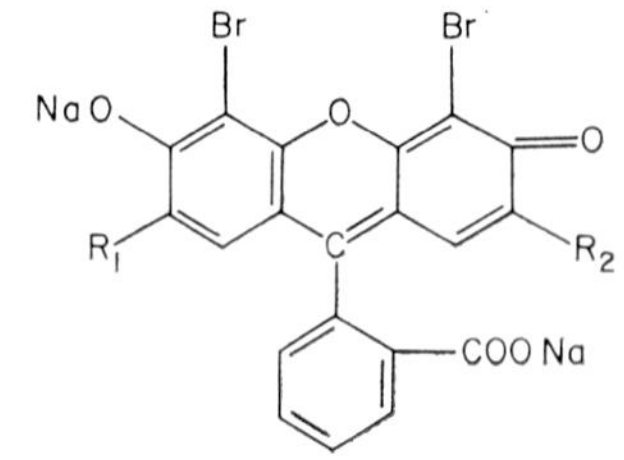

Fig. 1. Chemical structure of eosin. The two negative charges at the Na-binding sites. R_1, R_2 = Br:eosinY; R_1, R_2 = NO_2:eosin B.

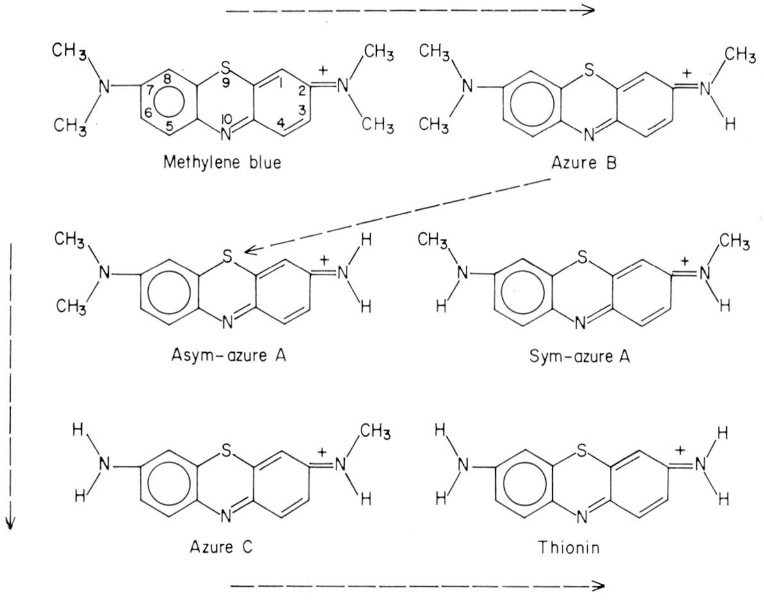

Fig. 2. Oxidative demethylation from methylene blue via the azures to thionin. Sym-dimethylthionin = sym-azure A apparently is outside this pathway. Its synthesis is not difficult; its staining properties are very similar to those of asym-azure A.

attempted to place the Romanowsky dyeing on a scientific basis. A problem which was disturbing at that time—and still is today—was the lack of reproducibility of results obtained by staining with random mixtures of polychromed Mbl and eosin. Giemsa appreciated the importance of standardization of the staining method and realized that it depended primarily on the standardization of the dyes used.

Giemsa showed that the red colour of parasitic protozoa nuclei and the brilliant purple colour of leucocytic chromatin was due to the combined action of eosin and of methylene azure I (= azure B) and that

Mbl was not required to obtain this nuclear colour. He established that, in the absence of eosin, azure B stained nuclei with a faint and dull. purplish hue, so that eosin was an indispensable component to enable the RD to stain cell nuclei deep purple. Giemsa added Mbl to the azure-eosin essentially to increase contrast between purple stained nuclei and the cytoplasm which stained greyish rather than pure blue with azure-eosin alone.

The question arises as to how pure Giemsa's azure B (methylene azure I) actually was. When azure A in really pure form became available in quantities sufficient to stain larger series of blood and bone marrow cells (Löhr *et al.*, 1974) it could be shown that under certain conditions cytoplasm can be stained blue solely with this azure. The present authors are convinced that a better understanding of these thiazine-eosin mixtures is indispensable for proper control of RD staining which so frequently differs from one commercial dye sample to the next, apparently due to "mysterious" vagaries (Cramer *et al.*, 1973) inherent in RD mixtures.

II. Features of Romanowsky Dye Solutions

RD are complex mixtures of cationic and anionic stains. Among the cationic components are Mbl and the azures, mainly azure B and A, but azure C and even thionin can also be detected in commercial products (Fig. 3). Equally, anionic eosin is hardly ever homogenous (Horobin and Murgatroyd, 1967; Marshall and Lewis, 1974; Marshall, Bentley and Lewis, 1974). By chromatographic separation, several fractions can easily be shown in different brands. The deep purple staining of nuclear chromatin is probably due to the interaction of the cationic azure molecules and anionic eosin molecules with DNA.

The question arises whether the staining features of RD are due to a combined action of several thiazine cations with eosin components or whether these complex "Romanowsky mixtures" can be replaced by one specific dye cation interacting with one dye anion. Elucidation of the molecular mechanisms involved in the RD may come from the studies by Caspersson *et al.* (1969, 1970) on chromosome banding with the dicationic fluorochrome quinacrine, and from the application of Giemsa dyes to investigate the banding phenomenon (Pardue and Gall, 1970). According to Summer and Evans (1971) and Summer *et al.* (1973), Giemsa banding occurs when a molecule of eosin approaches a region of DNA where two thiazine molecules are intercalated between base pairs at the correct distance apart, combining with both through its two acidic groups. A "magenta" coloured complex is formed with absorption maximum between 540 and 550 nm. This complex cannot be separated from the chromosomes without losing its absorption characteristics.

The Giemsa banding technique has also been extended to the study of whole human lymphocyte nuclei (Ross and Gormley, 1973), but analogous effects of RD and Giemsa on banding of metaphase chromosomes has not yet been intensively investigated. It is clear, however, that

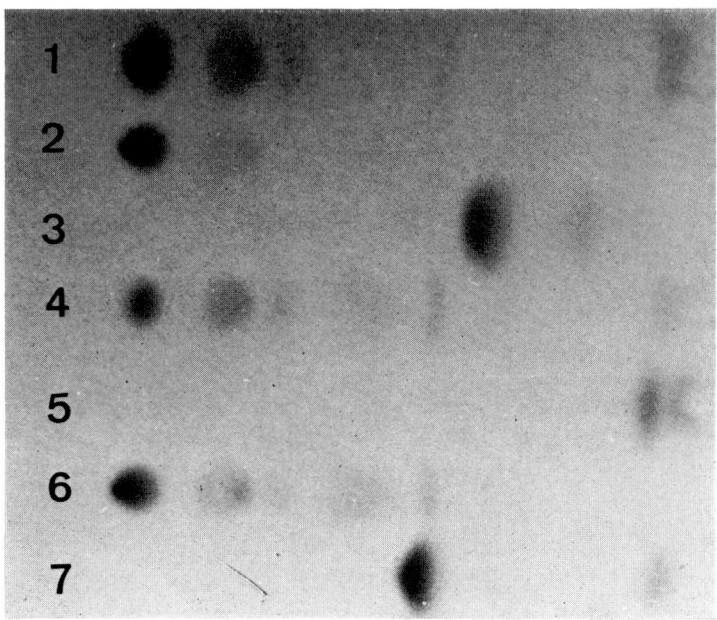

FIG. 3. TLC: Some thiazines and RD from various sources. (1) "May–Grünwald–Giemsa–Lösung": at least five fractions of thiazine dyes; (2) "methylenblau": azure B fraction on the right; (3) "thionin": fraction probably thionin on the left; (4) "Giemsa's azure–eosin–methylenblau": five fractions of Mbl, azures and thionin; (5) "erythrosin": at least two fractions; (6) "azure II": at least five fractions; (7) "azure C (purified)—eosin".

chromosomes and interphase nuclei are good models for studies on the purification of RD and the staining potential of their component compounds.

III. PURIFICATION OF ROMANOWSKY DYES

The thiazine-eosin mixtures which are included generally as RD are sold under a variety of different names—Wright's stain, Leishman's stain, May–Grünwald Giemsa's stain and also commercial samples of dyes labelled azure B, azure A and azure C. It is now well-recognized that no component of RD is sold in really pure form (see, for example,

Horobin and Murgatroyd, 1967; Horobin, 1969; Toepfer, 1972; Löhr and Wittekind, 1973; Marshall and Lewis, 1974). This applies especially to the azures where flagrant contradictions are found between inscriptions on labels and the content of the bottles. Thus, e.g. Fig. 4

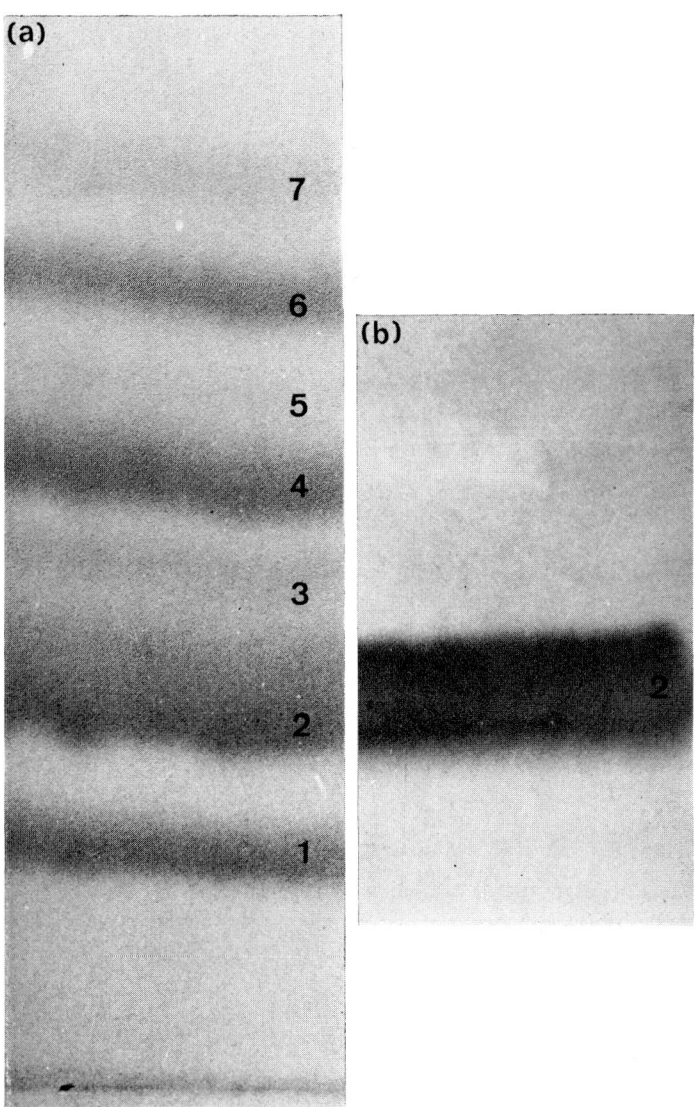

FIG. 4.(a) TLC: commercial samples labelled azure A "reinst" (purissimum). There are seven fractions: (1) Mbl, (2) azure B, (3) X_1 (4) azure A, (5) X_2, (6) azure C, (7) thionin. $X_{1,2}$:unidentified fractions. (b) TLC: azure B (2), purified from the sample (a) by column chromatography. Only very faint traces of methylene blue (and azure A) can be seen.

illustrates a commercial sample of so-called "reinst" (very pure) azure A separated by thin-layer chromatography (TLC) into several fractions from Mbl to thionin. In reality it is polychrome (oxidized) Mbl.

The implications of this state of commercial thiazines are obvious. Can standardization be attained by prescribing the exact amount of each thiazine component to be present in the mixture and by fixing the optimal relation between the total of thiazine cations to eosin anions? This has been proposed repeatedly in the past (e.g. Roe, Wilcox and Lille, 1941; Conn and Lillie, 1969). However, application of TLC has shown that not one thiazine dye constituent of RD is available in pure form, but each is invariably contaminated by other azures, while Mbl as sold always contains at least one additional fraction of azure B. Thus, nothing will be achieved from adding weighed amounts of commercial Mbl, azures and eosin to a solvent to arrive at a "standardized" RD. Mixtures have been mixed with mixtures, resulting in so-called "super mixtures" of the Romanowsky dyes, which vary in composition from one batch to the next and from one supplier to another.

A. THIAZINE-EOSINATES VERSUS ROMANOWSKY DYES

The solution of the problem must be looked for in another direction. Whether RD can be replaced by a single thiazine-eosinate and if so, by which thiazine, depends on the availability of the azures in pure form and in sufficient quantity to run large numbers of biological tests with blood and bone marrow smears to assess optimal staining conditions by varying dye concentration, staining time and addition of buffer solutions. Reliable separation methods are required in order to study the individual action of each azure. By thin-layer chromatography (TLC) and column chromatography (CC), it is possible to separate dyes such as the azures which are chemically so closely related that they may defy attempts at isolation by conventional methods of organic chemistry. Spectrophotometric assay of thiazine and other synthetic dyes supplement chromatography (Toepfer, 1970).

Among initial records of individual azure dye effectiveness were reports on specific interactions between living cells and the three azures (Löhr, 1972; Löhr and Wittekind, 1973; Wittekind et al., 1974). Our experience with air-dried blood films fixed in methanol confirms the individuality of each azure as a dye. When results obtained with pure dyes are consistent with those obtained with dye mixtures, it may be concluded that in the mixture one dye (e.g. azure A) may constitute the main fraction dominating the outcome of the staining procedure. What percentage of one or more "subfractions" of the other azures, Mbl and thionine can be tolerated in a mixture without perceptibly influencing the final staining result is a problem still to be elucidated.

It should be remembered that in our aim of standardization two factors must be considered. (1) The staining characteristics of each component need to be assessed in order to select the one most suitable to replace, together with eosin, the whole RD mixture. (2) Having recognized the "optimal" thiazine, methods have to be devised and applied to procure this dye in constant quality and sufficient quantity.

B. PURIFICATION OF EOSIN

As compared with the isolation of the thiazine components from RD, purification of eosin apparently is of relatively less importance, and there is no clear evidence that optimal staining of blood films depends on using highly purified eosin, eosin Y is apparently superior to eosin B. Nevertheless, a careful study on the influence of purified eosins on haematological staining seems appropriate.

IV. THIAZINE-EOSINATES AS SUBSTITUTES FOR ROMANOWSKY DYES

Roe, Wilcox and Lillie (1941) stated that the Romanowsky effect was given by all the azures, but especially by azure B; it was given faintly by toluidine blue and not at all by thionine. Although the authors used mixtures, their results are essentially consistent with ours obtained with pure azures and eosin (D. Wittekind and V. Kretschmer, unpublished). It seems likely that azure B eosin or azure A with eosin might be acceptable as a substitute for RD. Azure B-eosin solution can more easily be handled than solutions containing azure A, and it has the advantage that it is superior in demonstrating minute differences in cytoplasmic basophilia within the same cell. The cytoplasm of blast cells rich in RNA stains clear blue. Giemsa's original poor result with azure B-eosin, with the cytoplasm staining greyish rather than blue, seems likely to have occurred because his azure B was contaminated by azure A and, especially, by azure C. Our results indicate that addition of Mbl for correction of this deficiency can be dispensed with (Wittckind Löhr, Kretschmer and Sohmer, in preparation). A disadvantage of azure A-eosin is the difficulty in obtaining exactly reproducible staining of basophil cytoplasm from one purified sample of azure A to the next. Colour shades of cytoplasm and of cytoplasmic granules vary remarkably, apparently dependent on very exact balancing of the amount of azure A in relation to eosin in the staining solution. Excess eosin prevents nuclear staining by precipitation of azure A with eosin. Probably the absence of Mbl and of its stabilizing influence in other thiazine dyes in solution (Baker, 1958) is a factor. Another factor is the reduced solubility of azure A in aqueous solutions as compared with Mbl and azure B. However, we have obtained staining results which compare favourably with high quality RD by mixing azure A and eosin, each dissolved in methanol, and by adding water buffered to pH 6·4.

Azure C-eosin is a poor substitute for RD. Staining of nuclei is good, but the greyish red or purple cytoplasm lacks contrast with the nuclei. Nor can much be expected from toluidine blue which is itself a mixture of several thiazine fractions (Ball and Jackson, 1953; Di Berardino, 1954).

V. Availability of Pure Azure-eosins as Stains in Haematology

Introduction of a two-component haematological stain (one thiazine-eosin) necessitates availability of at least 500 mg of the thiazine. This fairly large quantity is required for assessing: (1) optimal relation of azure/eosin in stock and staining solutions; (2) optimal concentration of azure and eosin in methanol and in staining solutions; (3) use of suitable buffers in staining solutions; (4) staining time; (5) constancy of staining results assayed on a multitude of blood and bone marrow smears and tissue imprints.

Column chromatography and also solvent extraction (Bonneau *et al.*, 1967; Marshall and Lewis, 1975) are suitable methods for obtaining azures in adequate quantities. A more complex method is based on a combination of adsorption and ion-exchange chromatography (Löhr *et al.*, 1974).

VI. Quality Control of Thiazine-eosin Dyes

Careful consideration was given to this important subject by Lillie as long ago as 1944, when he found that in equal volume mixtures of methanol and glyercol, progressive alteration occurred of Mbl, azure B and azure A to lower azures. Since then there has been surprisingly little work on this important subject. Two questions are particularly pertinent. (1) What can be done to preserve quality of thiazine-eosin dyes? (2) What methods are suitable for controlling their quality?

A. PRESERVATION OF DYE QUALITY

Cationic thiazine dyes are stable in acid media only, and storage in alkaline media should be avoided. Some glassware may be a source of alkali, thus accelerating disintegration of thiazine dyes. Caps on bottles should not contain metal since various metals have the property of decomposing methylthionins (Lillie, 1946). Apolar solvents may be preferable as stock solutions, but azure chlorhydrates are stable for many months even in aqueous solution.

The optimal condition for preserving thiazine dyes probably is in the dry state as a chloride, sulphate or other anionic salt. However, thiazine salts, especially the chlorides, are very hygroscopic and extraction of the last traces of water may lead to destruction of the dye molecule (Lillie, 1944b). So far, reliable data on stability of azureeosinates is lacking.

Mbl and its demethylated homologues are known to be sensitive to daylight; storage should be in dark brown bottles (Cramer et al., 1973).

B. QUALITY CONTROL OF THIAZONE-EOSINS

Among the techniques suitable for checking the quality, especially of the azure, TLC seems to be of outstanding importance. Other methods of control include visible, i.e. NMR, spectroscopy and biological reaction. The purpose of such studies is to determine the percentage of coloured impurities observable in TLC which are required to produce changes in absorption characteristics and in biological staining. Consideration must also be given to the influence of salts, detergents and other materials which may be present, the effects of which are not yet finally known.

In conclusion, it is clear that Romanowsky stains are subject to considerable variation in their composition; stability and inter-related reactivity. It is hardly surprising that these variables influence the staining performance. There is need for a standardized stain and for quality control to ensure consistent staining of blood and bone marrow preparations. This paper describes the problems and indicates the way in which they are being resolved.

REFERENCES

Baker, J. R. (1958). "Principles of Biological Microtechnique". Methuen, London.
Ball, J. and Jackson, D. S. (1953). Stain Technol. 28, 33–40.
Bonneau, R., Faure, J. and Joussot-Dubien, J. (1967). Talanta 14, 121–122.
Caspersson, T., Zech, L. and Johansson, C. (1970). Expl Cell Res. 62, 490–492.
Caspersson, T., Zech, L., Modest, E. J., Foley, G. E., Wagh, U. and Simonsson, E. (1969). Expl Cell Res. 58, 141–152.
Conn, H. J. and Lillie, R. D. (1969). "Biological Stains", 8th edn. The Williams and Wilkins Company, Baltimore.
Cramer, A. D., Rogers, E. R., Parker, J. W. and Lukes, R. J. (1973). Am. J. clin. Path. 60, 148–156.
Di Berardino, M. (1954). Stain Technol. 29, 253–256.
Giemsa, G. (1902). Zentbl. Bakt. ParasitKde, Abt. I, Orig. 32, 307–313.
Giemsa, G. (1904). Zentbl. Bazt. ParasitKde, Abt. I, Orig. 37, 308–311.

Giemsa, G. (1907). *Dt. med. Wscher.* **33**, 676–679.

Harms, H. (1965). "Handbuch der Farbstoffe für die Mikroskopie". Staufen-Verlag, Kamp-Lintfort.

Horobin, R. W. and Murgatroyd, L. B. (1957). *Histochemie* **11**, 141–151.

Horobin, R. W. (1969). *Histochem. J.* **1**, 231–265.

Lillie, R. D. (1944a). *Publ. Hlth Rep.* **178**, suppl., 1–16.

Lillie, R. D. (1944b). *Publ. Hlth Rep.* **178**, suppl., 17–12.

Lillie, R. D. (1946). *J. Lab. clin. Med.* **31**, 253–256.

Löhr, W. (1972). Experimentelle Untersuchungen zur chromatographischen Reinigung und zur Vitalfärbung mit kationischen Thiazinderivaten. Inaugural-Dissertation, Med. Fak. Univ. Freiburg i. Brsg.

Löhr, W., Sohmer, I. and Wittekind, D. (1974). *Stain Technol.* **49**, 359–366.

Löhr, W., Sohmer, I., Wittekind, D. and Grubhofer, N. (1974). In preparation.

Löhr, W. and Wittekind, D. (1973). *Z. Zellforsch.* **137**, 125–140.

Marshall, P. N., Bentley, S. A. and Lewis, S. M. (1975). *Stain Technol.* (in press).

Marshall, P. N. and Lewis, S. M. (1974). *Stain Technol.* **49**, 351–358.

Marshall, P. N. and Lewis, S. M. (1975). (in press).

Nocht, B. (1898). *Zentbl. Bakt. ParasitKde, Abt. I, Orig.* **24**, 839–843.

Pardue, M. L. and Gall, J. G. (1970). *Science, N.Y.* **168**, 1356–1358.

Roe, M. A., Wilcox, A. and Lillie, R. D. (1941). *Publ. Hlth Rep.* **56**, 1906–1909.

Romanowsky, D. L. (1891). Oh the question of parsitology and therapy of malaria. Diss. Nr. 38 (in Russian), Imp. med. Milit. Acad. St Petersburg.

Ross, A. and Gormley, I. (1973). *Expl. Cell Res.* **79**, 81–86.

Summer, A. T. and Evans, H. J. (1973). *Expl Cell Res.* **81**, 223–236.

Summer, A. T., Evans, H. J. and Buckland, R. A. (1973). *Expl Cell Res.* **81**, 214–222.

Toepfer, K. H. (1970). *In* "Progress in Histochemistry and Cytochemistry 1", Nr. 5, Gustav Fischer, Stuttgart-Portland.

Toepfer, K. H. (1972). *Verh. dt. Ges. inn. Med.* **78**, 272–276.

Wittekind, D., Snipes, R. L. and Kretschmer, V. (1974). *Expl Cell Res.* **84**, 143–152.

13. Standardization and Quality Control of Anticoagulant Control

L. POLLER and G. I. C. INGRAM

The National Reference Laboratory for Anticoagulant Control,*
Withington Hospital, Manchester, M20 8LR; and Department of
Haematology, St Thomas' Hospital and Medical School,
London SE1 7EH, England

I. THE NEED FOR STANDARDIZATION IN ANTICOAGULANT CONTROL

Oral anticoagulants have a firmly established place in the management of venous thrombosis and the prophylaxis of thrombo-embolism. They are also used in some forms of arterial thrombosis. In disorders where anticoagulants are of benefit, clinical success depends on appropriate laboratory control. Unfortunately, there is variety in laboratory tests and in systems of reporting results, so that results at different hospitals may not be comparable.

* WHO Collaborating Centre for Anticoagulant Control Reagents.

The test which has been used in the majority of centres throughout the world is the Quick's one-stage prothrombin time, in which tissue extract ("thromboplastin") is used to accelerate the clotting of recalcified plasma. Nevertheless the Quick technique has been considerably modified since its original description (Quick *et al.*, 1935), and differences in the reactivity of various thromboplastins are the main cause of discrepancy between results at different hospitals.

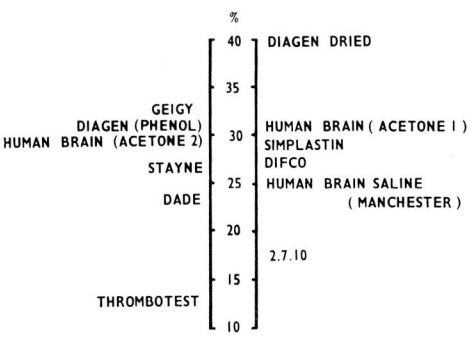

Fig. 1. Prothrombin activities derived from saline dilution curves. Mean of 30 patients.

This point is illustrated in Fig. 1 which shows the varying results obtained with different thromboplastins, all of which use the same basic technique and have been widely used in hospitals for the control of anticoagulant treatment. The results in Fig. 1 are reported as prothrombin activity from saline dilution curves made from normal plasma. Each of the results shown is an average based on 30 coumarin patient samples. Mean results range from 14–40% prothrombin activity. Most centres have used a therapeutic range of 15–30% prothrombin activity irrespective of the technique. It will thus be appreciated that in some hospitals there has been a tendency to under-anticoagulate and in others to over-dose with anticoagulant drugs simply because a different reagent has been used in the laboratory.

The use of prothrombin ratio, i.e.

$$\frac{\text{patient's prothrombin time}}{\text{normal prothrombin time}} \left(\text{e.g. } \frac{24}{12} = 2 \right)$$

has been recommended as an alternative. The use of the ratio has an advantage over prothrombin activity as it does not require the construction of dilution curves which are difficult to reproduce. To some extent it also controls technical variables (Biggs, 1965).

The substitution of prothrombin ratios for activities is however no

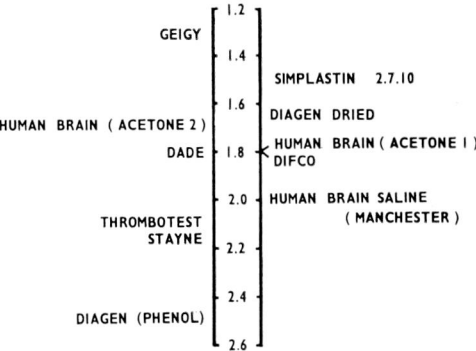

FIG. 2. The data of Fig. 1 as ratios. Prothrombin ratio = patient's prothrombin time/ normal prothrombin time.

easy solution to the problem of standardizing results as Fig. 2 shows. The results of the same 30 patients in Fig. 1 have been expressed as ratios. The problem is further illustrated in Fig. 3 where one plasma specimen has been tested with 15 widely used thromboplastin extracts.

The alternative expression, prothrombin index, has been employed in

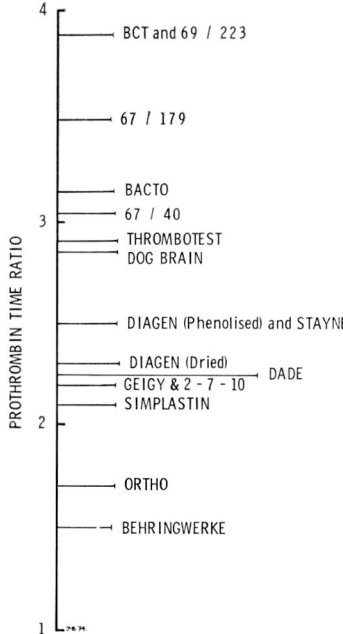

FIG. 3. Prothrombin ratios of a single test plasma using different thromboplastin extracts.

some parts of the world. This is also expressed as a percentage and has thus created added difficulties.

$$\text{Index} = \frac{\text{normal prothrombin time}}{\text{patient's prothrombin time}} \quad \left(\text{e.g. } \frac{12}{24} = 50\% \right).$$

The therapeutic limits with the prothrombin index are different from therapeutic percentages of prothrombin activity.

II. The British System

This chaotic state of laboratory control of anticoagulants existed in Britain before a system of standardization was initiated. The scheme began in Manchester in 1961 and by 1969 it had evolved into the British System for Anticoagulant Control. It began as a local routine supply scheme with all hospitals sharing a common thromboplastin reagent in their prothrombin time tests. Gradually it spread throughout the country. In 1969 the Manchester Comparative Reagent (MCR) was redesignated the national reagent, the British Comparative Thromboplastin. In addition, a national system of reporting, the British Corrected Ratio, was introduced so that all hospitals would have a common basis for reporting prothrombin results and for anti-coagulant dosage. The new ratio was to be given with every patient's results either alone or as an extra method of reporting if for any reason the hospital preferred to continue to use a different system of reporting (e.g. time, activity, index, etc.).

III. Recent Developments in Britain

Since 1969 further developments have occurred. At that time the majority of hospitals employed the BCT to calibrate their own home-made or commercial product in the "standardization procedure" described elsewhere (Thomson and Chart, 1970; Poller, 1970). In this way the equivalent prothrombin activities, ratios or indices with the local product which corresponded with the limits of therapeutic range with the Manchester reagent, were determined. This depended on a comparison of the results of the local reagent and the BCT when tested in parallel on coumarin-treated patients and fresh normal controls. With the greater availability of the BCT an increasing number of hospitals have begun to use the Manchester material for routine work. As a result comparatively few hospitals now use the BCT for reference purposes and the parent substance, the Manchester Comparative Reagent, has become the routine working reagent for many hospitals. As the result obtained is a direct ratio derived from the patient's prothrombin time divided by the normal value, the term British Ratio is used. The term British Corrected Ratio is retained for the result in a hospital in the reference scheme and is the ratio which would have been obtained had the BCT or MCR been used instead of the local thromboplastin reagent. The more widespread use of the standardized preparation for routine work is supported by a series of national quality control trials conducted in Britain using lyophilized plasma samples.

Having established the system on a national basis it seemed important to determine how accurately it was being used. There was a fundamental assumption that hospitals would obtain similar results with a uniform technique of the prothrombin time test and the use of the BCT. A continuous programme of quality control exercises was thus instigated in which all the hospitals using the Manchester thromboplastin participate. Hospitals are asked to test lyophilized plasma samples which are prepared in the National Reference Laboratory using the BCT.

Individual results are processed by computer and each hospital is provided with a printout (Fig. 4), giving the prothrombin time ratio of each plasma and the mean prothrombin time ratio at the individual hospital, the mean prothrombin time ratio and coefficient of variation for all hospitals together with the "per cent deviations" (the differences between the individual hospital's ratio and the mean expressed as a percentage of the latter). Figure 5 gives an example of the overall average deviation in one study. From this hospitals are able to assess their own performance in comparison with the other participants in the trial.

Analyses of variance were carried out to determine how much of the variability of the results obtained for each plasma was due to the tendency for some hospitals to obtain consistently different ratios. The aim was to assess the extent of variation between hospitals in measurements which should be identical, that is, those relating to a single plasma.

Results have confirmed that there is always considerable variation in the results reported when different hospitals try to measure the prothrombin time ratio of the same plasma with the standard thromboplastin (Leck et al., 1974). In the trials involving three plasma samples a marked tendency was demonstrated for hospitals which obtained a particularly high or low reading for one plasma to deviate in the same direction when measuring other plasmas in the same trial. However, this tendency to produce high or low readings in *one* trial was not consistent when the same hospital's results from *different* trials were compared in studies several months later. In other words, hospitals' tendencies to consistently overestimate the prothrombin time are

```
               NATIONAL REFERENCE LABORATORY
          FOR ANTICOAGULANT CONTROL REAGENTS

     QC TRIAL JUNE 1973
     256 LABS PARTICIPATED
     YOUR LAB NO 509B


     SPECIMEN       YOUR        ALL LABS        YOUR PER CENT
      NUMBER       RESULT      MEAN   C.V.       DEVIATION

     73/01          2.36       2.09   14.8          12.8
     73/02          1.75       1.80   13.3          -2.7
     73/03          3.02       3.33   17.6          -9.1
                   AVERAGE OF YOUR DEVIATIONS   8.3
```

FIG. 4. Sample of feedback to hospitals from the computer, comparing the recipient hospital's results with those of all hospitals in the same exercise. (Leck *et al.*, 1974).

```
HOW DO YOU COMPARE WITH OTHER LABS

   AV.DEV.
   0.1- 1.0
   1.1- 2.0   XXXXXXXXXXXXXXX
   2.1- 3.0   XXXXXXXXXXXX
   3.1- 4.0   XXXXXXXXXXXXXXXXXXXXXXXXXXXXXXXXXXXXXX
   4.1- 5.0   XXXXXXXXXXXXXXXXXX
   5.1- 6.0   XXXXXXXXXXXXXXXXXXXXXXXXXXXXXXXXXX
   6.1- 7.0   XXXXXXXXXXXXXXXXXX
   7.1- 8.0   XXXXXXXXXXXXXXXXX
   8.1- 9.0   XXXXXXXXXXXXXXX
   9.1-10.0   XXXXXXXXXXXXXXXXXXXXX
  10.1-11.0   XXXXXXXXXXXXX
  11.1-12.0   XXXXXXXXXX
  12.1-13.0   XXXXXXXX
  13.1-14.0   XXXXX
  14.1-15.0   XXXXX
  15.1-16.0   XX
  16.1-17.0   XXXXXXXX
  17.1-18.0   XX
  18.1-19.0   XXXX
  19.1-20.0   XX
  20.1-21.0   XX
  21.1-22.0   XXXX
  22.1-23.0
  23.1-24.0   XXXX
  24.1-25.0   XX
  25.1-26.0   XX
  26.1-27.0   X
  27.1-28.0
  28.1-29.0
  29.1-30.0   X
  30.0+       XXXXXXXX
```

FIG. 5. Sample of feedback to hospitals from the computer showing the distribution of mean percentage deviations of 256 hospitals. (Leck *et al.*, 1974).

largely short-term ones. Results obtained by the participants in the routine supply scheme were better than those of the national reference scheme hospitals suggesting that the regular routine use of the standard thromboplastin improves performance. The findings support the trend to reduction in the scale of the reference scheme in favour of routine supply of standardized thromboplastin reagent to British hospitals.

Confidence limits estimated from the results of all trials suggest that in the routine supply scheme the chances of having a true prothrombin

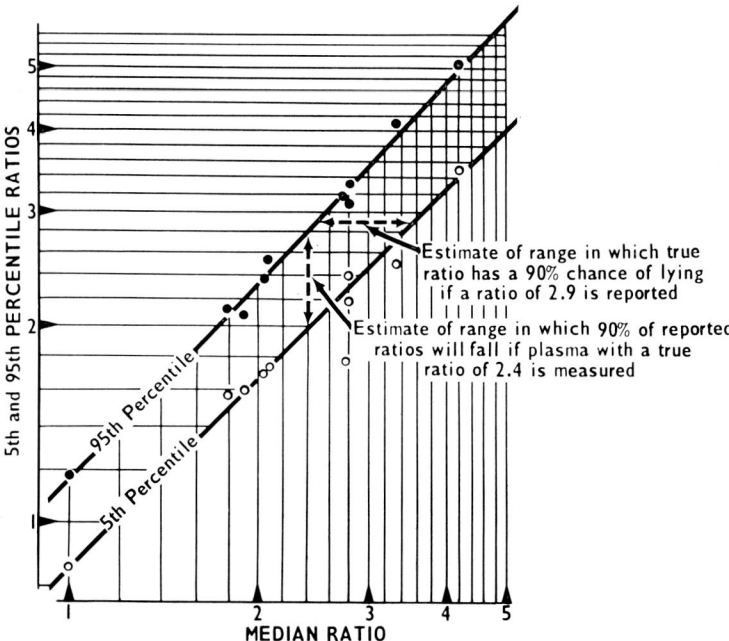

FIG. 6. Confidence limits for prothrombin time ratios obtained with BCT by routine supply scheme hospitals. (Leck *et al.*, 1974).

ratio outside the conventional therapeutic range (1·8–3·0) when the recorded ratio is 2·1–2·45 is only 5%. The point is illustrated in Fig. 6. The estimated fifth percentile ratio for each plasma is plotted as an open circle and the 95th percentile as a solid circle against the median ratio for the same plasma. Straight lines have been fitted by eye to the two sets of points. From the diagram it is possible to predict the range within which 90% of individual ratios might be expected to fall if a plasma of known true ratio were examined by a number of experienced laboratories. As an illustration, for a true ratio of 2·4 the limits are approximately 2·0 and 2·8 since these are the vertical co-ordinates of the points on the two diagonal lines which have a horizontal co-ordinate of 2·4. Figure 6 may also be used for predicting the range within which there is a 90% chance of the true ratio being located. Although these findings show the need for caution, they do not suggest that anti-coagulant dosage based on the data from the studies involves any special dangers.

One function of the national quality control trials is to stimulate self-criticism by enabling hospitals to assess their performance by relating their own results to the national mean and distribution of the results for all the participant hospitals. The best results were obtained from a select group of monitoring laboratories from which the technical staff have attended courses of instruction in the prothrombin time technique. Further improvement in laboratory performance may thus be achieved by an extension of such special programmes of practical training in prothrombin time technique.

IV. Technical Methods in the Preparation of British Comparative Thromboplastin

The standardized thromboplastin prepared at Withington Hospital, Manchester, is issued under two different labels:

(a) *Manchester Comparative Reagent* (MCR). This is used in the national routine supply scheme.

(b) *British Comparative Thromboplastin* (BCT). Once every 3–6 months a single batch of MCR is subjected to independent monitoring and subsequently designated the British Comparative Thromboplastin. It is then certified as conforming to the control scheme of the British Committee for Standards in Haematology. The details of the monitoring procedure and its independent assessment are given later.

The methods of production aim at the preparation of material which is identical between batches. Each batch of the routine supply MCR is matched against the BCT at the production centre so that it gives indentical results both with normal controls and with coumarin patients. The details are given of the steps required to produce the reference preparation, BCT. Most of the technical details of production also apply to the production of MCR, the routine reagent.

The BCT is a liquid phenol saline extract of human brain. The principal advantage of this preparation over a dry reagent is that it is easy to produce a large bulk of homogeneous material. Each batch of BCT is in current use for 6 months. It is despatched at monthly intervals to the hospitals in the reference scheme. BCT is stable for at least this period when stored at 4°C but is less stable at higher temperatures or if grossly contaminated in laboratory use.

As well as lending itself to large scale production in bulk and being homogeneous, the possibility of reconstitution errors is eliminated with a liquid extract. From the technical standpoint a large volume of a phenol saline extract is far less costly and time-consuming to prepare than an acetone-dried or freeze-dried preparation. Each batch of the BCT has to correspond to the previous batch. Uniformity between batches of BCT is ensured by the mass production of saline extract according to strictly controlled methods. Quality control of the material is performed at each step of production using lyophilized, banked laboratory standards stored at low temperatures.

After at least 2 weeks' storage at 4°C each brain is individually screened for gross insensitivity to the coumarin-induced clotting defect and for blood or bacterial contamination. The individual brains are then "pooled". Comparison of a new test "pool" is based on parallel results with previously approved batches of BCT. Graded saline dilutions of the test thromboplastin are prepared, i.e. neat, 80, 70 and 50% (see Table I). The results of these indicate, in this instance, that

TABLE I. Screening procedure for a new "pool"

Plasma dilutions (%)	Old BCT 44A	Old BCT 45B	Thromboplastins			
			BCT 46 (test)			
			Neat	80%	70%	50%
100	12	12	12	12	12	12·5
50	16	16	17·5	16·5	16	16
25	25	25	28	26	25	22
10	59	58	76	63	58	48

The entries are prothrombin times in seconds.

the 70% dilution of the "pool" is the closest to the previously approved batches, on the basis of a four-point saline dilution curve of normal blood.

In Table II the results of the second type of screening against coumarin plasmas are given. Here again the same impression is obtained, i.e. the 70% dilution of this small "pool" is the closest equivalent. Lyophilized factor-VII-deficient plasma ($<1\%$ activity) is also used in the screening. A large single batch of factor-VII-deficient plasma has been lyophilized at Withington Hospital for use as reference material in the National Reference Laboratory for

TABLE II. Screening against coumarin plasmas

| | Thromboplastins | | | | | |
Plasmas	Old BCT 44A	Old BCT 45B	Neat	BCT 46 (test) 80%	70%	50%
Fresh coumarin samples	25·5	25·5	29	27	25·5	24·5
	27	26·5	29	27·5	27	24
	31·5	31·5	36·5	34	32	31
	36	37	37·5	37·5	36	34·5
Lyophilized factor VII-deficient	35	35	38	37	35	34

The entries are prothrombin times in seconds.

Anticoagulant Control Reagents. It has been laid down in vacuum-sealed vials in a specially designated batch at −65°C. Degradation studies suggest that it is absolutely stable and should last for many years.

A new potential BCT "pool" must give identical results to the previously standardized batches within the limits of the experimental error of the method on both coumarin and normal plasma samples. At least five fresh normals collected under careful conditions (Poller, 1970) are included and it is preferable to use the same five persons for testing each "pool".

If the new material appears to satisfy the accepted criteria, a more elaborate procedure is next used with a ten-point saline dilution curve and parallel prothrombin times on at least 12 coumarin plasmas in comparison with previously accepted batches of BCT.

THE USE OF LYOPHILIZED REFERENCE MATERIAL IN PRODUCTION OF BCT

Progress has been made in the production of reliable lyophilized materials in recent years. It has thus been possible to incorporate both lyophilized thromboplastin and lyophilized plasma preparations in production methods for the BCT.

Prior to 1967 no reliable stable lyophilized thromboplastin reagent had been produced. Only since that date has it been possible to have any confidence in the results. Even now it would perhaps be unwise to place great reliance on a given lyophilized standard remaining stable indefinitely.

Some lyophilized thromboplastin reagents appear, from degradation studies, to be reasonably stable and thus play a valuable supportive role in BCT manufacture. Our first satisfactory internal lyophilized thromboplastin reagent was prepared in Manchester in 1967 and in the same year the Division of Biological Standards at Mill Hill, London, produced the first trial international reference preparation, designated 67/40. Both these preparations have proved adequately stable but both are in short supply so that further lyophilized preparations have been prepared as secondary reference material calibrated by the primary reagent. By 1970 five further preparations had been made at the National Institute for Biological Standards and Control, London, including two made from Manchester Reagent, one of which (69/223) appears to be reasonably stable. A large batch (sufficient for several years) of another lyophilized material (preparation 32) was also laid down at −65°C in the National Reference Laboratory for Anticoagulant Control Reagents, Manchester, as a secondary internal laboratory reference preparation for the production of BCT.

The current batch of BCT is checked regularly at the production centre with all the currently available primary and secondary lyophilized reference materials. This provides an extra safeguard against "long-term" drift of the BCT as well as monitoring the stability of the lyophilized thromboplastins. Each current batch of BCT, during its use in clinical practice is also monitored monthly at the production centre against lyophilized preparation 32 on at least 50 coumarin-treated patients.

Lyophilized plasmas play an important secondary role in BCT production. New batches of BCT must give a prothrombin time of 11–12 s with our lyophilized normal plasma. As explained earlier successive batches of BCT must also show no appreciable difference with the factor-VII-deficient reference plasma.

V. INDEPENDENT MONITORING OF SUCCESSIVE BATCHES OF BRITISH COMPARATIVE THROMBOPLASTIN

The independent monitoring system for BCT is concerned with the relative sensitivity of successive batches to the anticoagulant defect in the plasma of treated patients. Successive batches are very similar, so that the simplest method would to make a direct comparison of clotting times obtained with different batches on a series of patients' plasmas. However, BCT is intended for calibrating the sensitivity of other thromboplastins to the anticoagulant defect by determining equivalent ratios of prothrombin times with patients' plasmas to the times obtained with normal plasmas, so that similar ratios are used in the monitoring procedure for BCT itself.

If two batches of BCT have identical sensitivities to the anticoagulant defect, they would give the same ratio between the prothrombin times of a patient's and a normal plasma; or, as would be done in monitoring procedures, between the mean times from several patients and several normal plasmas. This may be expressed by the simple equation

$$P_1/N_1 = P_2/N_2 \qquad (1)$$

where N and P are the mean prothrombin times with the normal and the patients' plasmas, and 1 and 2 represent the results with the two thromboplastins. This may be written

$$(P_1/N_1)/(P_2/N_2) = 1 \qquad (2)$$

and in testing thromboplastins, departures from unity will indicate differences in sensitivity between them. The equation (2) may also be expressed as

$$(N_1/N_2)/(P_1/P_2) = 1 \qquad (3)$$

so that the principle comparison which is made for any two batches of BCT is between the average ratio of clotting times for normal plasmas and the average for patients' plasmas.

If the thromboplastins are identical,

$$N_1/N_2 = P_1/P_2 = 1. \qquad (4)$$

With BCT, successive batches of the same product are being monitored, so that both N_1/N_2 and P_1/P_2 would be expected to differ little from unity; the data are therefore examined for deviations here also.

To obtain measures of variability, several normal and patients' plasmas are tested and the ratio of the clotting times for two batches of thromboplastin are obtained from each plasma. Variation about the mean ratios for normal and the patients' plasmas is then calculated. For statistical convenience, all these comparisons are carried out on a logarithmic scale (Hills and Ingram, 1973).

The data for this analysis are obtained in the following manner. Twelve of the larger British laboratories undertaking anticoagulant control have been designated as monitoring centres, and have sent staff to the National Reference Laboratory for Anticoagulant Control Reagents to be trained in standard techniques. In rotation, groups of four monitoring centres receive from the National Reference Laboratory coded samples of the current batch of BCT, the proposed new batch and another batch. The provision of three batches makes it more difficult to guess from the results which is which, and is thus a safeguard against introducing unintentional bias into the results. The centres are asked to perform prothrombin times in duplicate with each sample on five normal plasmas and plasmas from six or more patients judged to be in the therapeutic range of oral anticoagulation, and to return their readings for analysis.

The quantity of data required on a given occasion has been determined from a retrospective study of data obtained in 1971 (Hills and Ingram, 1973). Average values were calculated for the standard deviations of the ratios of the clotting times for different batches of BCT with normal and with patients' plasmas. It was found that the values from normal and patients' plasmas were very close. However, when batches of material with perceptibly different sensitivities were compared, the mean s.d. was about double that obtained for comparisons between batches with very similar sensitivities. Taking the higher of these values, it was possible to calculate that at least four normal and six patients' plasmas would be required to detect the critical difference in sensitivity between two batches of BCT with an acceptable degree of certainty; and this quantity of data is required from at least three monitoring centres as a precaution against local variation: samples of the thromboplastins are therefore sent to four centres in case any one is unable to perform the tests on that occasion. Each monitoring exercise thus produces about 200 prothrombin time readings. The maximum permissible difference in sensitivity between successive batches of BCT was of course a matter of clinical judgement; and it was considered that if the ratio of a given patient's plasmas with a current batch of BCT were 2·0, it would be adequate for anticoagulant control to obtain values lying between 1·9 and 2·1 with other batches. It has therefore been the practice to accept a new batch of BCT provided that the 95% confidence limits for the ratio with the new batch lie within the range 1·9–2·1 for a ratio of 2·0 with the reference batch.

This system is administered by the Anticoagulant Control Panel of the British Committee for Standards in Haematology. The panel is most grateful for the very careful work of the monitoring centres, which makes it possible to obtain results with this high degree of precision. Figure 7 gives a visual impression of the precision obtained, by plotting means of duplicate readings as log. prothrombin times for a known insensitive batch of thromboplastin (called "weak") and for a proposed new batch, against the current batch, designated "old". The lines are drawn at 45° in both instances; the difference between the "weak" and "old" batches is apparent, and it may be verified that the log. deviations among the normal and patients' readings are of the same order. The close similarity between the results obtained with the "new" and "old" batches reflects the quality of the standardization procedures employed at the National Reference Laboratory, before the batches of BCT reach the monitoring procedure.

So far, monitoring has used individual, fresh plasma samples from normal subjects and from treated patients, as would be used in a working laboratory to calibrate another thromboplastin against BCT. The results seem satisfactory but the method is laborious. Furthermore, only wet samples of thromboplastin are tested, which have a limited shelf-life. The panel is now introducing experimentally a parallel series of tests with lyophilized normal and patients' plasmas, and to

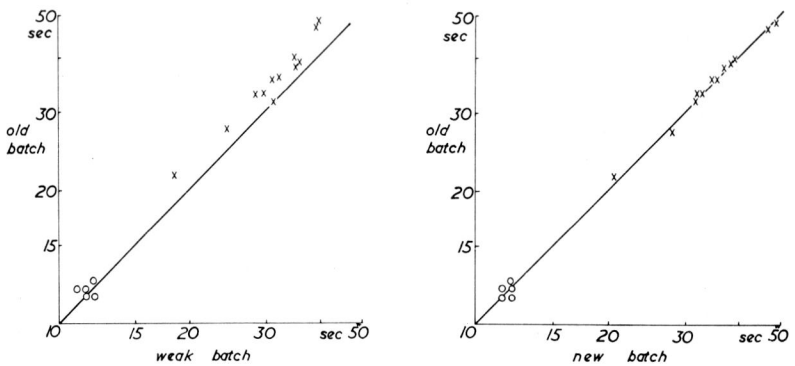

FIG. 7. Monitoring of British comparative thromboplastin. Means of duplicate prothrombin times on 12 patients on oral anticoagulants (×) and five normal subjects (○) with three batches of thromboplastin. A proposed "new" batch and a known insensitive ("weak") batch were compared with the previous ("old") batch. Both lines are drawn at 45°. The "weak" batch is easily detected.

compare wet samples of BCT with the similar reference preparation (69/223) held lyophilized at the National Institute for Biological Standards and Control in London, and with a lyophilized reference thromboplastin prepared at the National Reference Laboratory.

VI. Approaches to Standardization in Other Countries

We have described the British system which basically depends on a nationally approved thromboplastin reagent, a national system of reporting, and proficiency quality control trials with lyophilized plasma samples at a proportion of individual hospitals. Other investigators have suggested that standardization might be achieved in different ways, e.g. by using freeze-dried reference plasmas alone or with various freeze-dried reference thromboplastin preparations.

A. REFERENCE STANDARD PLASMAS

Miale (Miale and LaFond, 1969; Miale and Kent, 1972; 1973) has prepared artificial plasmas, produced commercially by Warner Chilcott,

U.S.A. ("Verify") i.e. normal plasma and two artificial coagulation defects intended to correspond to different degrees of coumarin treatment. Lyophilized plasmas are also available from Nyegaard, Oslo, Norway, which represent three levels of normal plasma activity (10, 25 and 50%) with added fibrinogen and factor V. The latter are claimed to be inactivated, suitable for quality control within laboratories and for revealing possible defects in reagents, techniques and equipment. Loeliger has prepared for study lyophilized pooled plasmas from treated patients with different degrees of anticoagulation (Bangham *et al.*, 1973).

The protagonists of reference plasmas claim that they might be used in several ways. A thromboplastin could be standardized against reference plasmas by adjusting its sensitivity to give prothrombin times within certain limits on each of two (or more) reference plasmas, or to give a ratio, between certain limits, of one prothrombin time to the other. Alternatively, reference plasmas might be used to calibrate thromboplastins, so that laboratories could calculate the prothrombin time (or prothrombin time ratio) which would have been obtained with an "ideal" thromboplastin, from the results obtained with a different but calibrated thromboplastin. A different use of reference plasmas would be to calibrate a given test: suppose two plasmas were available which represented the limits of the therapeutic range for patients; then a test could be run on each of these plasmas in a given batch of clinical samples, to classify the patients into those below, within and above the therapeutic range, irrespective of the actual clotting times obtained on a given day. Again, reference plasmas could be used in quality control of prothrombin time testing at individual hospitals: samples could be included in runs of clinical tests, and results within certain limits would show that the system was working dependably; samples tested day by day could also be used to detect temporal drifts.

There are, however, a number of problems to be overcome before reference plasmas can fulfil these various roles; they must be shown to be stable, reproducible from batch to batch, and vial to vial. They must give satisfactory end-points in both manual and automated tests. If used to characterize different thromboplastins they must not be appreciably activated in the course of preparation, because some reagents would then give misleadingly short prothrombin times. Also, some thromboplastins are more sensitive than others to the inhibitory effects of PIVKA(s), the abnormal forms of clotting factors released during anticoagulant treatment. The clinical significance of PIVKA(s) is not yet clear but it may be important to use a thromboplastin which takes this inhibitory activity into account; hence reference plasmas should ideally copy clinical samples in this respect. Clearly those made from patients' plasmas would be more likely to do so than those made artificially.

B. LYOPHILIZED REFERENCE THROMBOPLASTINS

Another approach is based on the assumption that stable lyophilized reference preparations of thromboplastins of different types can be manufactured, analogous to biological standards in other fields. The use of such materials in the production of BCT has already been discussed. Reference thromboplastins could then be given arbitrary indices showing their relative sensitivities to the anticoagulant defect, determined by comparative tests between the various preparations on a panel of plasmas from treated patients, or perhaps against reference plasmas. A series of five lyophilized thromboplastins is held by the National Institute for Biological Standards and Control, London (Bangham *et al.*, 1973), corresponding to various types of human and animal reagents commonly used clinically in anticoagulant control. Batch 69/223 is prepared from Manchester Comparative Reagent; 67/40 is human brain with bovine factor V and fibrinogen; 70/115 is rabbit brain with bovine factor V, fibrinogen and cephalin; 70/178 is phenolized rabbit brain; 68/434 is bovine brain, factor V and fibrinogen with cephalin.

Problems have been encountered with some reference thromboplastins, particularly in their behaviour in clotting machine determinations. The National Insitute series of thromboplastins have shown good stability in accelerated degradation tests. Nevertheless the literature does not yet suggest that these thromboplastin preparations have been widely adopted in anticoagulant control. They are in all events only to be used for characterization of other standardized or national working reagents. Their stability has not so far been subject to independent assessment although a programme involving the 69/223 preparation has recently been devised by the British Anticoagulant Panel.

C. ICSH INTERNATIONAL STUDY

In order to compare the behaviour of various reference plasmas against reference thromboplastins and to contrast these with the behaviour of the thromboplastins when tested on fresh clinical samples, the International Committee for Standardization in Hematology has established an expert panel on oral anticoagulant control to conduct an international study. Experiments are being undertaken from which it is hoped to obtain useful evidence on which to base national policies for reference materials which might be used in the standardization of anticoagulant control. The study is also designed to give evidence on equivalent therapeutic ranges by various techniques and with various reagents.

D. THE INTERNATIONAL STUDY GROUP FOR ANTICOAGULANT CONTROL

Meanwhile, the British system may be found to be a useful model for the organization of a national scheme for uniform anticoagulant control, with modifications appropriate to different national situations. The International Study Group for Anticoagulant Control is co-operating with colleagues in 38 countries to promote the development of such national schemes.

Although no specific procedures or reagents are recommended at this stage, the benefit of the cumulative experience of all, including those involved in the ICSH Study and in the British System, is thus becoming generally available and may provide a basis for the international organization of anticoagulant control. Conclusions on standards reached by current or future clinical trials may then be usefully applied in all countries.

ACKNOWLEDGMENT

We are grateful to the Editor and Publisher of the British Journal of Haematology for permission to reproduce Figs 4, 5 and 6.

REFERENCES

Bangham, D. R., Biggs, R., Brozović, M. and Denson, K. W. E. (1970). In "Vascular Factors and Thrombosis" (F. Koller, K. M. Brinkhous, R. Biggs, N. F. Rodman and S. Hinnom, eds). F. K. Schattauer Verlag, Stuttgart.

Bangham, D. R., Biggs, R., Brozović, M. and Denson, K. W. E. (1973). *Thromb. Diath. haemorrh.* **29,** 228–239.

Biggs, R. (1965). *Thromb. Diath. haemorrh.* Suppl. **17,** 304–360.

Hills, M. and Ingram, G. I. C. (1973). *Br. J. Haemat.* **25,** 445–451.

Leck, I., Gowland, E. and Poller, L. (1974). *Br. J. Haemat.* **28,** 601–612.

Leck, I., Thomson, J. M. and Poller, L. (1973). *Br. J. Haemat.* **25,** 453–460.

Miale, J. B. and Kent, J. W. (1972). *Am. J. clin. Path.* **57,** 80–88.

Miale, J. B. and Kent, J. W. (1973). *Am. J. clin. Path.* **60,** 453–457.

Miale, J. B. and LaFond, D. (1969). *Am. J. clin. Path.* **52,** 154–160.

Poller, L. (1970). Association of Clinical Pathologists Broadsheet No. 71.

Quick, A. J., Stanley Brown, M. and Bancroft, F. W. (1935). *Am. J. med. Sci.* **190,** 501–510.

Thomson, J. M. and Chart, I. S. (1970). *J. med. Lab. Technol.* **27,** 207–212.

14. Control of Diagnostic Haematology Products

ROBERT M. SCHMIDT

*Hematology Division, Center for Disease Control,
Public Health Service, U.S. Department of Health,
Education and Welfare, Atlanta, Georgia 30333, U.S.A.*

I. INTRODUCTION

The subject of this chapter, control of diagnostic haematology products, covers a broad area of laboratory medicine and encompasses governmental, professional, industrial, and individual standard-setting mechanisms and concepts of quality control. The organization of interested persons into standard-setting bodies is not new; however, these bodies are now putting an increasing emphasis on quality of product as well as test method because of the emergence of a growing diagnostic reagents and instruments market. Indeed, few clinical laboratories can afford to evaluate diagnostic products and must rely upon the advice and reputation of the manufacturer and input by governmental or professional groups. Out of necessity most of this chapter deals with haematology diagnostic products marketed in the United States. In addition, the prevalence of laboratory haematology tests and their use in different types of facilities is emphasized as a basis for discussing the regulation of diagnostic products nationally and internationally.

II. THE ROLE OF THE CLINICAL LABORATORY IN HEALTH CARE DELIVERY SYSTEMS

Health care delivery has been estimated to be the second largest industry in the United States, and is exceeded only by defence. The current increased demand for improved health services, especially in the area of preventive medicine, is largely due to education and the emergence of third party payment of health care delivery. Multiphasic screening clinics, routine hospital in-patient work-ups, pre-school anaemia screening programmes, maternal and child health clinics, and genetic disease screening programmes emphasize laboratory techniques as an adjunct to patient history and physical examination. In the United States over $83 billion was expended on health care (7·6% of the gross national product) in 1972 (U.S. Department of Commerce, 1973).

Since the organization of the first medical laboratory in the United States at the Johns Hopkins Hospital in 1889, laboratory medicine has grown into a major industry. In October 1971, the first national census of clinical and public health laboratories was conducted, and 12 295 clinical laboratories were reported (Table I). However this figure does not include laboratories in doctors' offices where there are fewer than five physicians, and it is estimated that at least 4000 more laboratories are in this category.

Increasingly, laboratory tests are being used to detect and diagnose disease and potential disease states. Approximately 3 billion *in vitro* diagnostic tests are currently performed in the United States each year with an annual increase of 15% predicted (First, 1972). The retail dollar value of clinical laboratory services for 1970 has been estimated at $1·3 billion with a projection of $2·8 billion for 1980 (Jacobson, 1973). Thus, the clinical laboratory represents a major business enterprise in the United States, attracting much attention from multiple sectors of society.

Before it is possible to discuss the impact of diagnostic products on

TABLE I. Clinical laboratories in the United States 1971*

Clinical laboratory	Number	Total
Nongovernment		
Hospitals	4069	
Outpatient clinics	431	
Private (independent)	2922	
Group practice	1293	
Industrial	127	
Other	165	
		9007
Government		
State-local		
Hospitals	2100	
Outpatient clinics	85	
Health Departments	392	
Other	119	
		2696
Federal		
Hospitals	447	
Outpatient clinics	108	
Other	37	
		592
Total		12 295

* Clinical and Public Health Laboratory Survey 1972. U.S. Department of Health, Education and Welfare, Public Health Service, Center for Disease Control, Atlanta, Georgia.

TABLE II. Performance of clinical laboratory services in the United States 1970 and 1980*

	Doctor office laboratories (%)	Small neighborhood laboratories (%)	Hospital laboratories (%)	Large independent laboratories (%)	Total (%)
1970	39	28	13	20	100
1980	12	28	13	47	100

* Reprinted with permission from Medical Marketing and Media, pp. 1–8, February 1973.

haematology, it is necessary to evaluate the current source of laboratory services and the estimated change in the next few years. Jacobson (1973) has shown that most laboratory tests are presently being performed in doctors' offices (Table II). By 1980, however, large independent laboratories are expected to provide 47% of the laboratory services. Automation in large-test facilities will become increasingly important, thereby altering the market for current diagnostic products.

Thus far this discussion has been limited to laboratory medicine in the United States. Few hard facts on international laboratory medicine exist. However, Prak and Fizette (1974) presented an overview of the diagnostic product industry at an International Conference on Standardization of Diagnostic Materials co-sponsored by the World Health Organization (WHO) and the Center for Disease Control (CDC) in June 1973. The diagnostic reagent industry is probably composed of between 1000 and 1300 companies which produce over 100 000 different products that are used in clinical laboratories. The industry is concentrated in the United States, Japan, Germany, United Kingdom and several other western European countries. Prak and Fizette believe that few products are currently being made under fully documented quality control procedures although numerous products have some elements of quality control in their manufacturing procedure.

No exact figure exists for the international expenditure for diagnostic products. Jacobson (1973) estimates that nearly $240 million was spent for diagnostic products in the United States in 1971.

III. The Haematology Laboratory in the United States

The Robert S. First Corporation (1972), a market research organization, estimates that 846 million laboratory tests in general haematology were performed in the United States in 1972 (Table III). Hospital laboratories performed 65·5% of the tests, independent laboratories performed 24·8%, and group practice and physicians' offices performed 12·7% of haemoglobin determinations but only 7·0% of overall general haematology laboratory tests (in contrast to Jacobson's estimate for overall laboratory tests).

Ninety-seven million coagulation tests were performed in the United States in 1972 (Table IV). These tests were primarily done by hospital laboratories (82·0%). Only a smaller number were performed by independent laboratories (9·3%).

In immunohaematology, hospital blood banks perform the vast majority of tests (62·1%) followed by Red Cross blood centres (21·1%).

TABLE III. Haematology testing volume by facility—1972 (in millions)*

Test	Facility Hospital	Independent labs	Group practice	Physician's office	Total tests
Hb	120	58	8	27	213
PCV	100	36	3	5	144
WBC	110	45	5	10	170
RBC	90	27	1	2	120
Differential leucocyte count	95	31	3	7	136
Platelet count	21	4	–	1	26
Other	18	9	3	7	37
Total (No./%)	554 (65·5%)	210 (24·8%)	23 (2·7%)	59 (7·0%)	846

* Reprinted with permission from A United States Study of Diagnostic Reagents and Systems (1972). Robert S. First, Inc., 405 Lexington Avenue, New York, N.Y. 10017.

TABLE IV. Coagulation testing volume by facility—1972 (in millions)*

Test	Facility Hospital	Independent labs	Group practice	Physician's office	Total tests
P.T.	59·2	6·5	1·1	3·8	70·6
P.T.T.	12·2	1·0	0·1	0·3	13·6
Others	8·0	1·5	0·7	2·4	12·6
Total (No./%)	79·4 (82·0%)	9·0 (9·3%)	1·9 (2·0%)	6·5 (6·7%)	96·8

* Reprinted with permission from A United States Study of Diagnostic Reagents and Systems. (1972). Robert S. First, Inc., 405 Lexington Avenue, New York, N.Y. 10017.

Community, commercial, and military blood banks provide the remainder of immunohaematology laboratory services (Table V).

In the United States, haematology tests (specifically Hb determination) are the most commonly performed laboratory procedures. Less than 10% of these tests are performed in a physician's office.

Annual surveys of hospital and non-hospital laboratories offering haematology services are carried out by *Laboratory Management* (Table VI). Non-hospital groups and hospital groups each provide approximately 50% of the haematology services (including coagulation), but as

TABLE V. Immunohaematology testing volume by facility—1972 (in millions)*

			Facility			
	Hospital blood banks	Red Cross centres	Community	Commercial	Military	Total
ABO grouping	12·1	3·7	1·7	1·3	0·3	19·1
Anti-Rh_0 (D)	10·2	3·7	1·7	1·3	0·3	17·2
Other anti-Rh/Hr	3·6	6·5	0·9	0·3	0·2	11·5
Antibody screening	5·8	3·7	1·7	1·3	0·5	13·0
Antiglobulin (Coombs) direct	2·1	0·6	0·3	0·2	0·4	3·6
Antiglobulin (Coombs) indirect	12·4	0·1	0·1	—	0·4	13·0
Crossmatch	14·2	0·1	0·1	—	0·8	15·2
Other rare typing, etc.	1·3	0·3	0·3	0·2	0·2	2·3
Total anti-sera	61·7	18·7	6·7	4·6	3·1	94·9
Hepatitis associated antigen	4·2	3·7	1·7	1·3	0·3	11·2
Grand total (No./%)	65·9 (62·1%)	22·4 (21·1%)	8·5 (8·0%)	5·9 (5·6%)	3·4 (3·2%)	106·1

* Reprinted with permission from A United States Study of Diagnostic Reagents and Systems (1972). Robert S. First, Inc., 405 Lexington Avenue, New York, N.Y. 10017.

TABLE VI. Laboratories offering haematology services in the United States 1972–73

	Number of laboratories	
	Haematology	Immuno-haematology
Non-hospital laboratories*	6372	3662
Hospital laboratories†	6562	4444
Total	12 934	8106

* National survey of non-hospital clinical laboratories. *Laboratory Management* **12**, 30–51, 1974.

† Survey of hospital clinical laboratories. *Laboratory Management* **11**, 22–33, 1973.

expected, more hospital laboratories provide immunohaematology services than do the other facilities.

IV. HAEMATOLOGY DIAGNOSTIC PRODUCTS IN THE UNITED STATES

During January 1974 a letter was sent to the president of each firm in the United States which manufactures and/or markets diagnostic products. From the 461 letters sent out, 252 responses were received; 122 firms stated that they were involved with haematology diagnostic products. The survey is summarized in Table VII. Products were divided into six major categories: (A) blood collection products; (B) cell counting and whole blood measurements; (C) smears, differential counts, and stains; (D) abnormal haemoglobin detection products (D) coagulation products; (F) multiple purpose items. (Immunohaematology products were not included because these are already licensed by the Bureau of Biologics and monitored by that organization.)

This survey represents the first extensive identification of diagnostic haematology products marketed and or manufactured in the United States. Three hundred and eight products were identified including a diverse collection of haematology reagents, kits, instruments and ancillary equipment (blood collection tubes, syringes, multiple purpose laboratory items).

This type of listing is necessary for three major reasons: (1) to be used in consultations with laboratory directors on the utility of specific products; (2) to serve as a source list of laboratory products which can be used by both consumer and manufacturer; (3) to serve as a basis of the development of product class standards (see section V).

The names and addresses of the firms represented on this list are available at no cost (Center for Disease Control, 1974).* Complete product files with descriptions of products available have been developed at the CDC. This information is also available to anyone upon request.* Finally, information obtained from this survey may serve as a nucleus of product information for ongoing discussions of the development of standards of performance for these groups of products. The number of products listed in Table VII indicates that this must be a long-term commitment of the United States government and will be a formidable responsibility. The CDC is currently working with professional groups, industrial representatives, and the Food and Drug Administration in developing this programme.

Although haematology tests represent a large volume of requested

* From Hematology Division, CDC, Atlanta, Georgia, 30333, U.S.A.

laboratory procedures, haematology diagnostic products have a relatively small sales volume and command a small share of the market when overall diagnostic products are considered. For example, only three haematology products (Ortho's Anti-Human Coombs Test, Coulter's Lysing Agent, and Coulter's Isoton) are listed in the ten leading diagnostic products marketed in the United States (Jacobson, 1973). Many of the "best sellers" are *in vivo* diagnostic materials including radiochemicals and contrast media. Thus although haematology tests are the most commonly performed laboratory procedures, they account for a disproportionately small share of the total diagnostic reagent expenditure in the United States.

New diagnostic products are developed through several mechanisms. (1) Industry develops a new product for a laboratory test which saves time and money, or is well-adapted to special situations, for example, a glucose screening test. (2) Industry develops a market for a new product. For example, Technicon SMA-12 resulted in a greater use of blood chemistry in managing patients, which in turn resulted in new clinical research. (3) Industry responds to a public health programme by developing new product lines. For example, the National Sickle Cell Disease Program brought about the development of such products as Sickledex. (4) Physicians and researchers demand new tests because of new clinical and research findings. For example, the development of associated antigen tests now used in immunohaematology laboratories and a test for hexoseaminidase A deficiency (Tay-Sach's disease) for use in genetic screening programmes.

The CDC will continue to monitor haematology diagnostic products as they appear on the market and will work with other governmental, professional, industrial, and private organizations to ensure quality of product.

V. Haematology Kits

Two investigators have written extensively about medical usefulness, accuracy, precision specificity and sensitivity of various commercial kits for clinical chemistry (Barnett, 1965, 1968, 1972; Logan, 1972). More recently similar papers have appeared evaluating haematology kits in this manner (Schmidt and Wilson, 1973; Schmidt and Holland, 1974; Wilson and Schmidt, 1974; Filip, Eckstein and Sibley, 1973). This topic is dealt with in chapter 15. Therefore, evaluation of commercial haematology kits is not emphasized in the present chapter, but several points about utility and the influence of kits on haematological technology are summarized.

The term kit was defined by an ad hoc committee which met at the

Center for Disease Control in Atlanta, Georgia, in 1969: "A *kit*, *reagent set* or *diagnostic aid* is a collection or assembly of reagents, devices, or equipment, or a combination thereof, offered for sale or distribution and containing all of the major components and written instructions necessary to perform one or more designated diagnostic tests or procedures."

In vitro diagnostic products, a product class, and a product class standard were defined in the Federal Register (1973): "*In vitro diagnostic products* are those reagents, instruments, and systems intended for use in the diagnosis of disease or in the determination of the state of health in order to cure, mitigate, treat, or prevent disease or its sequelae. Such products are intended for use in the collection, preparation, and examination of specimens taken from the human body. These products are drugs or devices ... (as defined in the Federal Food, Drug, and Cosmetic Act) ... or are a combination of drugs and devices, and may also be a biological product."

"A *product class* is all those products intended for use for a particular determination or for a related group of determinations or products with common or related characteristics or those intended for common or related uses. A class may be further divided into subclasses when appropriate."

"A *product class standard* is a statement describing performance requirements necessary to assure accuracy and reliability of results, specific labelling requirements necessary for the proper use of a particular class, and procedures for testing the product to assure its satisfactory performance."

The majority of haematology kits are used in small clinical laboratories in physicians' offices or hospitals, and in independent laboratories where this type of test is performed in small volume. Kits such as those for haemoglobin determinations and solubility tests for Hb S are marketed with the small laboratory in mind. They allow a quick evaluation of the patient without a special visit to a hospital or other laboratory, and do not necessitate the mailing of a blood sample to a central laboratory. Selection of kits by the physician or laboratory supervisor depends upon a number of factors summarized by Laessig (1973): (1) availability; (2) physician need; (3) test accuracy/ precision; (4) quality of personnel available to perform the test; (5) test instrument available or required; (6) medical usefulness; (7) cost.

Obvious problems include inability to choose from a muliplicity of kits, lack of specific training of personnel performing tests (office nurse, clerk, technician), and inadequate training and/or interest of supervisor or physician in laboratory medicine. Frequently quality control is deficient or non-existent, and these small laboratories are not licensed

TABLE VII. Haematology diagnostic products in the United States

Product	Number of manufacturers	Number of marketing firms
(A) Blood collection products		
(1) Blood collection tubes	6	16
(2) Capillary tubes	12	19
(3) Lancets	5	13
(4) Syringes	3	9
(5) Alcohol swabs	3	6
(6) Other swabs	1	5
(7) Arterial blood sampling kit	1	1
(8) Bone marrow biopsy tray	1	1
Subtotal	32	70
(B) Cell counting and whole blood measurements		
(1) Pipettes and diluters	23	32
(2) RBC and WBC electronic counting equipment	8	14
(3) Mechanical counting equipment	9	17
(4) Automatic platelet counting equipment	5	10
(5) Osmotic fragility testing devices	1	3
(6) Haematocrit measuring instruments	9	20
(7) Haemoglobinometers	9	16
(8) Haemoglobin reagents	17	25
controls	7	9
standards	7	10
(9) Foetal haemoglobin kits	3	5
(10) Haematology reference controls	7	8
(11) Erythrocyte sedimentation rate tubes	4	7
(12) Oximeters	2	2
(13) Photometers	4	6
(14) Gamma counters	2	2
(15) Blood volume measurement	1	1
Subtotal	118	187
(C) Smears, differential counts and stains		
(1) Automatic differential counters	4	8
(2) Blood stains and buffers	12	18
(3) Automatic stainers	4	8
(4) Cytochemical stains	5	9
(5) Slides, coverslips	8	9
(6) Automatic slide smearer	1	1
Subtotal	34	53

TABLE VII (*continued*)

Product	Number of manufacturers	Number of marketing firms
(D) Abnormal haemoglobin detection products		
(1) Solubility kits, instruments	9	12
(2) Haemoglobin electrophoresis kits, instruments	14	20
(3) Cellulose acetate membranes	8	11
(4) Cellulose acetate electrophoresis reagents	5	7
(5) Densitometers	11	16
(6) Abnormal haemoglobin controls	5	10
Subtotal	52	76
(E) Coagulation products		
(1) Coagulation timers	9	20
(2) Prothrombin time reagents and/or	10	16
(3) Partial thromboplastin time reagents and/or controls	6	10
(4) Platelet factor reagents	2	7
(5) Factor deficient plasmas	4	5
(6) Fibrinogen determination kits and/or reagents	3	9
(7) Fibrinogen and fibrin split products reagents and/or kits	5	5
(8) Fibrinogen	1	1
(9) Coagulation expendable items	3	3
(10) TGT testing	2	2
(11) Correction reagents	2	2
Subtotal	47	80
(F) Multiple purpose items		
(1) Specimen mailers	4	4
(2) Blood mixers	5	6
(3) Centrifuges	4	6
(4) Microscopes and accessories	4	7
(5) Timers	4	5
(6) Balances	4	4
Subtotal	25	32
Grand total	308	498

or regulated by any agency or professional society. Probably fewer than 10% of all clinical laboratory services are provided in laboratories large enough to evaluate commercial products. Thus, the consumer must depend upon the reliability of the kit manufacturer.

Two recent surveys have shown that physicians' offices perform a large variety of haematology laboratory tests (Gleich and Rose, 1973; Jackson *et al.*, 1973), including tests not usually considered routine for office practice such as WBC differentials, RBC counts, and pro-thrombin times. In a survey of limited size, it was also shown that laboratories in physicians' offices performed less reliably than patho-logist-directed laboratories in the determination of haemoglobins and PCV (Schoen *et al.*, 1971). It is difficult however to implicate any single cause–effect relationship. Commercial products are only one factor. The education and training of the personnel responsible for perfor-ming the tests also play an important role.

VI. Automated Instruments for Haematology

Automation was introduced into the clinical laboratory in 1957, shortly after the development of diagnostic kits. It is currently estimated that 80% of the approximately 15 000 clinical laboratories in the United States use some automated equipment (Frost and Sullivan, 1973). However, until recently this automation was primarily found in the clinical chemistry laboratory. The market for haematology analyses is now rapidly expanding, and it is expected to be a $72 million industry in the United States by 1980 (Fig. 1). In 1972, 479 haematology analysers turned out 210 million tests. Frost and Sullivan estimate test volume will increase fourfold by 1980.

Automated haematology instruments are now available for particle counting, haemoglobin, PCV, red-cell indices, prothrombin time, and partial thromboplastin time determinations, as well as for blood banking procedures and white blood cell differentials. Additionally, semi-automated equipment represents an enormous market in haemato-logy which is beyond the scope of this chapter. A newly emerging market includes equipment to monitor the on-line status of patients in intensive and coronary care units (e.g. blood gases). As more laboratory haematology tests are performed in regional rather than small labora-tories, automated haematology equipment can be expected to become the major source of haematology test results.

Automated instruments have alleviated several problems in labora-tory haematology, but, as expected, they have also produced a few new ones. Some of these are summarized in Table VIII. The use of auto-mated haematology analysers has not always resulted in a great savings

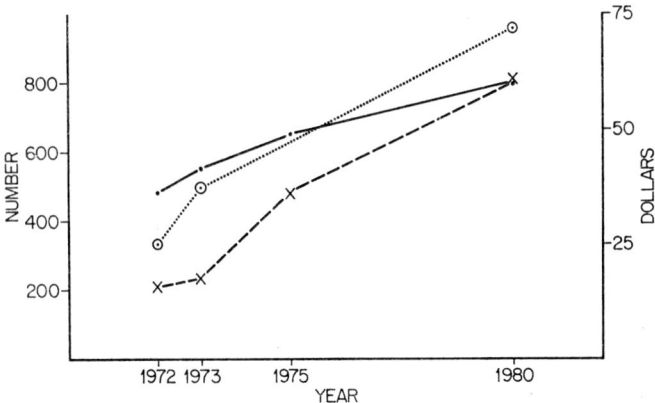

FIG. 1. Automated haematology instruments marketed in the United States. . ———.,
number of instruments; × – – – ×, test volume (in millions); ...., instrument sales (in
millions,—does not include optional equipment, reagents, supplies or significant overseas
sales of equipment). Data reprinted with permission from Automated Clinical Laboratories,
published by Frost & Sullivan, Inc., 106 Fulton Street, New York, N.Y. 10038, in May 1973.

to the patient. Examples have frequently been cited where a patient's
sample is tested with a multichannel analyser and an abnormal result is
obtained. An extensive chemical, and perhaps physical, examination
then ensues at great emotional and financial expense to the patient.
Laboratory technicians and directors as well as clinicians must be
educated in these new technologic advances to prevent abuse. On the
positive side, multiphasic screening tests have turned up unsuspected
disease states and have resulted in a decrease in morbidity and
mortality.

TABLE VIII. Automated instruments for haematology

Advantages	Disadvantages
(1) Lower average cost/test	(1) Unnecessary tests done by
(2) Rapid reliable results	multichannel analysers
(3) Allows hook-up to computer	(2) Physicians disregard addi-
(4) Fewer errors in information	tional tests
transfer	(3) Complex instruments must be
(5) Fewer technologists required	run by highly trained personnel
(6) Allows multiphasic screening	(4) Customer service erratic
at lower cost	(5) Requires regionalization of
(7) Allows regionalization of	laboratory services
laboratory services	

The laboratory director is currently faced with a multiplicity of automated haematology instruments and must select the one best suited for his needs. Mechanisms for insuring quality control of diagnostic products are discussed in the next section of this chapter. But one major problem, the promotion of new devices before they are ready to be marketed, requires emphasis here. Premarketing testing and procedures for quality control should be completed before the product is distributed. Few laboratorians can spend the time required to work out the problems of a new system. Additionally, patient lives are affected by such products, and the margin for error in diagnostic laboratory results should be minimal. The major manufacturers have long followed good manufacturing practices which should now be adopted by the entire diagnostic industry.

In the area of automated instrumentation, special equipment is being developed to perform the actual tests requested by physicians rather than a set of tests predetermined by the instrument manufacturer. Discrete haematology analysers should obtain a major share of the market and prevent the excessive flow of laboratory data which the physician neither needs nor requests. Reagents and instrument time are wasted when only one test is required. There is also a need for automated instruments capable of performing emergency procedures in intensive care units and emergency rooms. Finally, instruments should not limit a user to a given methodology or a required reagent system available only from the instrument manufacturer.

Automation of haematology laboratory tests coupled with computerization of test results will increasingly replace manual and semi-automated test procedures. The development of multiphasic screening centres, the emphasis on preventive medicine, third party payment of health laboratory costs, and development of regional or central laboratories all reinforce the need for standardization and control of automated instrumentation. American manufacturers are currently competing with their European and Japanese counterparts in marketing automated instruments abroad. Both developing countries and industrialized countries will provide this overseas market which, it is predicted, will exceed domestic sales before 1980.

VII. MECHANISMS FOR ENSURING QUALITY CONTROL OF HAEMATOLOGY DIAGNOSTIC PRODUCTS

Standardization of laboratory diagnostic products was the topic of a recent meeting held at the Center for Disease Control (CDC) in Atlanta, Georgia. The meeting was co-sponsored by the World Health Organization (WHO) and CDC. An eminent group of 65 scientists representing the disciplines of clinical chemistry, haematology, and microbiology made recommendations for the world-wide standardization of laboratory techniques and diagnostic products ("Proceedings of International Conference on Standardization of Diagnostic Materials", 1973). These recommendations form the basis of a programme in laboratory standardization adopted by the 1974 World Health Assembly. Although the actual mechanism of developing the new WHO programme is under discussion, it will probably include some mechanism of ensuring quality control of diagnostic products. Mechanisms for ensuring quality control fall into two main categories, voluntary and mandatory.

A. VOLUNTARY CONTROL

Voluntary approaches to quality control of diagnostic products include good manufacturing practices. Most firms have developed systems of quality control in the manufacturing of diagnostic products which follow basic sound management and scientific principles. This should result in built-in quality rather than elimination of unsatisfactory products at the end of the production cycle. Prak and Fizette (1974) discussed the impact of good manufacturing practices on diagnostic reagents at the WHO Standardization Conference in 1973. Objectives are well-known to all industrial manufacturing personnel and have been proposed as guide lines for diagnostic product quality control by the U.S. Food and Drug Administration (Federal Register, 1973).

Another voluntary approach to control of diagnostic products is the monitoring of products by professional, governmental, and private groups. For example, in the United States the College of American Pathologists (CAP) maintains a haemoglobin standards laboratory at the Cleveland Clinic. On a voluntary basis, manufacturers of cyanmethaemoglobin (hemiglobincyanide) standard solutions submit each batch to the CAP laboratory where testing is performed and the batches are certified (King and Willis, 1970).

The International Committee for Standardization in Hematology (ICSH) has developed an international hemiglobincyanide reference standard which is produced by the Institute of Health of the Netherlands under a grant from the Council of Europe (Van Assendelft, 1972). This solution is made available upon request to laboratories as a reference standard. Thus, it is possible for manufacturers to compare the properties of their reagents with those of an internationally accepted reference standard. In addition the ICSH has prepared international recommendations for standardization of specific haemotologic test procedures, including the erythrocyte sedimentation rate (ESR) and determination of serum iron which could be used as a basis for standard-izing diagnostic products. Twenty panels and committees of the ICSH are working on other standards which may subsequently be adopted by national or professional organizations to ensure quality laboratory practices in haematology.

The National Committee for Clinical Laboratory Standards (NCCLS has an area committee in haematology which has adopted a standard method for the erythrocyte sedimentation rate and is currently working on prothrombin time, nomenclature of blood cells, and particle coun-ters. This organization is uniquely composed of a group of representa-tives from government, industry, and professional societies. Currently 17 professional associations, seven governmental agencies, and 51 industrial organizations are members. Thus, standard writing involves all interested parties in the United States and a consensus mechanism is established on a voluntary basis.

The World Health Organization has designated international biological standards in the area of immunohaematology. Current standards include anti-A, anti-B and anti-Rh O (anti-D) blood typing sera, with others in the "proposed" stage (Engelfriet, 1974). With its developing laboratory standardization programme, WHO will clearly expand its role in designating reference standards.

A *voluntary programme* has several shortcomings. First, since not all manufacturers participate, reliable firms paying for quality control procedures may be placed at an economic disadvantage by competitors who do not use such procedures. Similarly, the consumer has no assurance of quality of products manufactured by nonparticipating firms. Second, a manufacturer may market a product found to be unsatisfactory (without an accreditation seal) since no regulatory mechanism exists. Third, if a product does not meet the requirements of one standard-setting body, the manufacturers can market the product in another country or area where that body has no jurisdiction or reputation. Fourth, voluntary controls tend to be permissive without a closely-scrutinized system of checks and balances, and they can con-fuse both the consumer and manufacturer.

B. MANDATORY CONTROL

Two countries have enacted mandatory control of diagnostic products. The Federal Republic of Germany has developed standards for methods and reagents which are published in the German Pharmacopoeia (Büttner, 1974). The United States is developing a regulatory mechan-ism for medical devices including diagnostic products through a new programme of the Food and Drug Administration. Details of this activity are given in section VIII.

Although the Council of Europe has proposed regulations governing the manufacture and marketing of certain preparations for the rapid diagnosis and monitoring of diseases through Resolution AP (70), regulator control is currently impossible (Büttner, 1974).

VIII. The Regulation of Diagnostic Haematology Products in the United States

The Food and Drug Administration (FDA) has had the authority to regulate diagnostic products used in the clinical laboratory since the enactment of the Food, Drug, and Cosmetic Act by congress in 1938. However, an exemption from labelling requirements for these products was provided for diagnostic reagents. In January 1972 the Commissioner of the Food and Drug Administration published a statement in the Federal Register (1972a) indicating that the FDA would regulate diagnostic products. This was followed by several Federal Register publications describing labelling requirements for diagnostic products and the designation of product class standards (Federal Register, 1972b, 1973, 1974a).

Currently two federal agencies are involved in the development of the national regulations for diagnostic products in haematology: the Center for Disease Control (CDC) in Atlanta and the Food and Drug Administration (FDA) in Washington, D.C. The major provision of the new regulations is specific labelling information upon which compliance action can be taken if a product fails to perform as stated by the manufacturer. Current activity has stressed the requirement of specific product labelling which were announced on 15 September 1974 (Federal Register, 1974a).

The major responsibility of the FDA Diagnostic Products Programme, however, is the establishment of product class standards whereby groups of products having common uses or components may be evaluated at one time. The CDC is serving as the major source of technical advice to the FDA and is responsible for preparing the initial drafts of these product class standards. Each product class standard should include the following detailed information: (1) definition and intended use (2) specific labelling information, including storage, stability, batch number, intended use, summary of methodology and principle of test, limitation of procedure, expected results, bibliography, and performance requirements (accuracy, precision, sensitivity, specificity); (3) performance requirements; (4) method of evaluation; (5) certification procedure.

The CDC and the FDA are working extensively with expert consultants drawn from universities, government, and industry. Working groups meet to propose the preliminary product class standard and to determine priorities. Four major areas are stressed in the development of product class standards. (1) A reference method; (2) a standard procedure (specifications); (3) reference materials; (4) quality control procedures.

Three proposals for product class standards in haematology have been written and presented to the FDA Diagnostic Products Advisory Board: (1) quantitative haemoglobin determination (Federal Register, 1974b); (2) turbidity tests for Hb S; (3) Hb electrophoresis kits.

Current priorities for product class standard development in haematology include: (1) prothrombin time; (2) partial thrombo-plastin time; (3) fibrinogen determination; (4) red-cell enzymes; (5) cell counting; (6) haematocrit (PCV); (7) erythrocyte sedimentation rate; (8) blood collection procedures. This is a long term programme, and several product classes will be published in the Federal Register each year.

The haematology diagnostic products regulatory programme can only be successful if input is received from both the consumer and manufacturer at an early stage. The CDC is working with the National Committee for Clinical Laboratory Standards (NCCLS) to ensure that all interested parties can review preliminary drafts prior to submission to the FDA. In addition, a product class standard published by the FDA only becomes valid after a variable "call" period during which interested parties may respond with suggestions. By working with these groups in the actual writing of the standard, the government hopes to minimize arbitrary decisions and maximize the current state of the art with provisions for future scientific and clinical achievement to be incorporated into the standard.

The FDA is charged with the legal responsibility of regulating the diagnostic products programme and is developing a network of surveillance and compliance to carry out the programme.

In addition to the FDA's Bureau of Medical Devices and Diagnostic Products, the Bureau of Biologics is also involved with regulating diagnostic products. Since 1947, the Bureau of Biologics has licensed products used in the immunohaematology laboratory. The authority for this activity is Section 351 of the U.S. Public Health Service Act which states that products which are used as an aid in diagnosis of disease and which are biological products or used with biological products must be licensed.

The Bureau of Biologics currently licenses three broad categories of immunohaematology products:

(1) red-cell grouping and typing serum; (2) reagent red cells; (3) Coombs' serum.

Proposed products for regulation include anti-HLA typing serum and transfer factor. A manufacturer initially applies for a license to produce an immunohaematology reagent and concurrently applies for a license to manufacture a specific product. The licensing activity involves three steps: (1) evaluation of production protocol and labelling of the product; (2) testing of samples at the Bureau of Biologics; (3) inspection of the plant. The plant must be inspected at least once each year, and any change in protocol must be submitted to the Bureau for approval. Each lot of reagent is certified by the Bureau, and a release form must be obtained before it can be distributed.

Although the current regulation of diagnostic products in the United States is based on the Food, Drug, and Cosmetic Act, a Medical Devices amendment was proposed by the Congress to allow regulation of all medical devices and products under one programme.

IX. INTERNATIONAL CONTROL OF DIAGNOSTIC PRODUCTS

The role of government, professional societies, and other groups in developing standards for haematology diagnostic products has been described in the preceeding eight sections of this chapter. It is obvious that no single mechanism for insuring control of haematology diagnostic products on an international basis currently exists. Although the ICSH has been concerned with developing standard test methods for haematologic procedures, this activity has not emphasized quality control o diagnostic products.

The recent mechanism for developing haematology product class standards in the United States could serve as a model for other industrialized countries. However, developing countries have needs in laboratory haematology which must be met by specially designed diagnostic products. These countries are largely dependent upon several major American and western European manufacturers for their diagnostic products. There is an acute shortage of trained personnel in these countries, and there are practically no scientists trained in quality control and laboratory medicine. Thus products which ultimately reach these countries are frequently used indiscriminately without evaluation. Additionally, several months may elapse from the time of shipping these products to their arrival. Products which may meet basic good manufacturing practices and national regulations when produced, may deteriorate in tropical climates or during prolonged shipment. There have also been reports of shipment to developing countries of immunohaematology products which are unacceptable for sale in the United States.

Adadevoh (1974) suggested the development of freeze-dried products and special packaging of products destined for export to remote areas. He also suggested that a surveillance system of quality control be undertaken by an international agency. A list of accredited products could then be made available to national agencies and individual laboratory scientists. Manufacturers would be encouraged to submit their products for accreditation.

It is obvious that international co-operation is required to ensure uniform quality of laboratory haematology products which meet the needs of both developed and developing countries. The World Health Organization is uniquely suited for this role as a central clearing house of product and standard reference information, and to co-ordinate the related activities of professional scientific and governmental bodies.

Use of trade names is for identification only and does not constitute endorsement by the Public Health Service or by the U.S. Department of Health, Education and Welfare.

REFERENCES

Adadevoh B. K. (1974). *In* "Proceedings of International Conference on Standardization of Diagnostic Materials", Atlanta, Ga., June 5–8, 1973, pp. 69–72. Center for Disease Control, Atlanta.

Barnett R. N. (1965). *Am. J. clin. Path.* **43**, 562–569.

Barnett R. N. (1968). *Am. J. clin. Path.* **50**, 671–676.

Barnett R. N. (1972). *In* "Progress in Clinical Pathology", (M. Stefanini, ed.) vol. 4, pp. 181–198. Grune & Stratton, New York and London.

Büttner, H. S. E. (1974). *In* "Proceedings of International Conference on Standardization of Diagnostic Materials", Atlanta, Ga., June 5–8, 1973, pp. 55–59. Center for Disease Control, Atlanta.

Center for Disease Control (1974). "Hematology Diagnostic Products in the United States". Center for Disease Control, Atlanta, Ga.

Engelfriet C. P. (1974). *In* "Proceedings of International Conference on Standardization of Diagnostic Materials", Atlanta, Ga., June 5–8, 1973, pp. 115–117. Center for Disease Control, Atlanta.

Federal Register (1972*a*), Vol. 37, 819.

Federal Register (1972*b*), Vol. 37, 16613–16617.

Federal Register (1973), Vol. 38, 7096–7102.

Federal Register (1974*a*), Vol. 39, 8610.

Federal Register (1974*b*), Vol. 39, 9217–9219.

Filip, D. J., Eckstein, J. D. and Sibley, C. A. (1974). *Am. J. clin. Path.* **62**, 32–39.

First, R. S., Inc. (1972). "A United States Study of Diagnostic Reagents and Systems". New York.

Frost and Sullivan, Inc. (1973). "Automated Clinical Laboratories". New York.

Gleich, C. S. and Rose, E. F. (1973). *Lab. Med.* **4**, 14–17.

Goddard, James L. (1973). *Scient. Am.* **229**, 161–166.

Jackson, J. A., Vastbinder, E. and Hamburg, J. (1973). *Lab. Med.* **4**, 17–20.

Jacobson, W. R. (1973). *Medical Marketing and Media*, pp. 1–8.

King, J. W. and Willis, C. E. (1970). *Am. J. clin. path.* **54**, 496–501.

Laessig, R. H. (1973). *Medical Lab.*, pp. 20–25.

Logan, J. E. (1972). *CRC Crit. Rev. clin. Lab. Sci.* **3,** 271–289.

Prak, J. L. and Fizette, S. B. (1974). *In* "Proceedings of International Conference on Standardization of Diagnostic Materials", Atlanta, Ga., June 5–8, 1973, pp. 61–67. Center for Disease Control, Atlanta.

"Proceedings of International Conference on Standardization Materials", Atlanta, Ga., June 5–8, 1973, Center for Disease Control, Atlanta.

Schmidt, R. M. and Wilson, S. M. (1973). *J. Am. med. Ass.* **225,** 1225–1230.

Schmidt, R. M. and Holland, S. (1974). *Clin. Chem.* **20,** 591–594.

Schoen, I., Thomas, G. D. and Lange, S. (1971). *Am. J. clin. Path.* **55,** 163–170.

U.S. Dept. of Commerce, Statistical Abstract of the United States (1973). Government Printing Office, (No. 0324-00108), Washington, D.C.

Van Assendelft, O. W. (1972). *In* "Modern Concepts in Hematology" (G. Izak and S. M. Lewis, eds) pp. 14–25. Academic Press, New York and London.

Wilson, S. M. and Schmidt, R. M. (1974). *Clin. Chem.* **20,** 1138–1140.

15. Product Evaluation

IRWIN M. WEISBROT

Norwalk Hospital, Connecticut 06856, U.S.A.

I. Introduction

A. DEFINITION

In its broadest sense product evaluation is a continuing function of a laboratory's quality surveillance programme, but in this discussion the term will be used in a limited sense to signify a formal evaluation, based on a well-defined experimental protocol designed to answer specific questions as to a product's performance. A brief discussion, and the necessarily limited experience of a single group, does not permit an all-inclusive view (ranging from qualitative "kits" for infectious mononucleosis testing to huge combined coagulation and counting instruments). The illustrative examples in this chapter are limited largely to diagnostic quantitative methods such as haemoglobinometry and particle counting.

B. ORGANIZATIONAL STRUCTURE OF PRODUCT EVALUATION

Although the number of organizations concerned directly or indirectly with the evaluation of diagnostic methods, "kits", devices or instrumental systems is incredibly large, there seems to be little coherent structure directed towards the actual physical task of evaluating products. Much effort is expended in defining terms, establishing principles, and setting standards. As product evaluation has commercial, academic and regulatory connotations, manufacturers, scientists and government officials must all become involved.

1. *ICSH*

The main direction of the ICSH has been to standardize methodology but the committee's continuing work in the surveillance of purified cyanmethaemoglobin solution is, in a sense, a form of product evaluation at the bench. A similar programme for an international certification of particles is in progress.

2. *WHO*

A major accomplishment relating to product evaluation was the International Conference on Standardization of Diagnostic Materials co-sponsored with the U.S.A. Center for Disease Control in June 1973.

3. *National organizations (U.S.A.)*

(a) *NCCLS*. The National Committee for Clinical Laboratory Standards is a voluntary committee, with representatives of scientific organizations, government and industry, whose main goal is to define and specify. Among their projects is the design of a statistical protocol for the evaluation of laboratory methods. Although this has not yet been used for an evaluation of a haematological product, some details are described below.

(b) *CAP*. The College of American Pathologists provides many scientific services useful in product evaluation. Among these the certification of cyanmethaemoglobin solutions is most germaine, but other clinical standard solutions useful in the laboratory are prepared. The data produced in the College proficiency surveys and its quality assurance service are a prime source of data concerning product and method performance.

The Product Evaluation Subcommittee (PES-C) of the CAP provides a service to manufacturers for the laboratory evaluation of their products. Its main purpose is to test advertising claims and provide for objective verification. Verified claims may be advertised as such under the college seal. Manufacturers are required to use discretion in such advertising and their products are subject to re-evaluation if subsequent field reports indicate failure to meet the verified claims.

(c) *HEW*. The Department of Health, Education and Welfare has proprietary and regulatory interests in product evaluation through its branches.

The Food and Drug Administration (FDA) has an *in vitro* Diagnostics Division. Requirements of product labelling include documentation precision, accuracy, specificity and sensitivity. The division may do some product evaluation in its own laboratory or contract for evaluations of selected products as it deems appropriate. The Division of Biologicals sets standards and does batch testing of products such as blood typing sera.

(d) *Other organizations.* The National Bureau of Standards provides standard reference materials, with a growing list of materials of interest to the clinical laboratory. The National Science Foundation has concerned itself with the problem of ethically combining manufacturers' needs for product evaluation of medical instruments and devices with those of the university scientists for free publication rights.

Similar organizations are established in other countries, for instance, in the U.K. the Department of Health and Social Security has a Laboratory Developments Advisory Group (LDAG) which is concerned with the quality and performance of laboratory equipment. The Department of National Health and Welfare of Canada is co-operation with the Canadian Standards Association (1973), has issued an excellent report describing the need for a specific funded council for product evaluation.

II. SOURCES OF PRODUCT EVALUATION INFORMATION

A. PUBLICATIONS

Most instrument systems have been the subject of recent published evaluations. An excellent review of haematological instruments was given by Brittin and Brecher (1971). Each worker has devised his own basis of comparison but almost all specifically confront the problems of precision and accuracy. Publications of the Commission on Continuing Education of the American Society of Clinical Pathologists are valuable. Its *Technical Improvement Service* reports provide short monographic evaluations of methods (and instruments), as do the *Check Sample* series. The *Summary Report* provides short anecdotal comments, useful at times, and always amusing. The broadsheets prepared by the Association of Clinical Pathologists (U.K.) are very good as are the reports and bulletins of the Association of Clinical Biochemists (U.K.).

B. ADVERTISEMENTS

These are often charming, but the rare statistical statements are usually in the superlative (rather than the comparative) and give rise to much scepticism.

C. PROFICIENCY SURVEYS

Valuable insight into product performance can be culled from properly reported and presented surveys. Tables I, II and III sum= marize the CAP data available in our files for the performance of representative methods and instruments for haemoglobinometry and cell counting. Although initial reports of performance were encouraging,

TABLE I. Performance of haemoglobin methods, CAP proficiency surveys 1971–73 (20 specimens)

	Manual HiCN	IL Haemoglobino- meter	Coulter model S	Fisher Hem-Alyser
$\bar{x}$ (g/dl)	11·19	10·96	11·06	11·05
Pooled CV (%)	3·85	5·24	2·48	3·16*

* Based on 14 available CV's.

TABLE II. Performance of red-cell counting methods, CAP proficiency surveys, 1971–73 (18 specimens)

	Chamber	Coulter model A, B, D, F	Coulter model S	Fisher Hem-Alyser	Technicon SMA-4/7
$\bar{x}$ ($\times 10^{12}$/l)	3·959	3·826	3·958	3·827	3·840
Pooled CV (%)	8·79	5·42	2·21	5·84*	6·23†

* Based on 10 available CV's.
† based on 12 available CV's.

TABLE III. Performance of white cell counting methods, CAP proficiency surveys 1971–73 (11 specimens)

	Chamber	Coulter model A, B, D, F,	Coulter model S	Fisher Hem-Alyser	Technicon SMA—4/7
$\bar{x}$ ($\times 10^9$/l)	19·28	19·62	18·21	17·55	16·49
Pooled CV (%)	14·33	8·08	7·12	—	18·42†
	(13·01)	(5·22)	(4·36)	(10·11)*	(17·35)

( ) CV calculated omitting one specimen, $1\cdot5 \times 10^9$/l, CV's *c.* 20% for all methods.
* Based on 8 available CV's.
† Based on 7 available CV's.

neither the IL Haemoglobinometer (Gambino and Waraksa, 1969) or the Technicon SMA 4/7 (Green *et al.*, 1969; Lappin *et al.*, 1969) has done well in the CAP surveys. One limitation of survey data may be inherent in the synthetic material provided. We have observed differences in performance of our laboratory's methods on different surveys. These differences may also lie with differing methods of statistical analysis. Uniformity of the statistics of proficiency surveys would be helpful.

The alarming variability of blood typing sera noted in the 1970 CAP proficiency survey by Koepke and Thakur (1972) was not noted in a follow up study of the 1971 survey (Koepke and Cichetti, 1973). It is not known if this represents improvement in the reagents or in the survey material. Survey data for simple coagulation tests indicate the need for better methods.

III. PROBLEMS IN HAEMATOLOGY PRODUCT EVALUATION

A. STANDARD REFERENCE MATERIALS

Product evaluation would be simplified if standard reference materials were universally available.

1. *Haemoglobin*

The availability of certified cyanmethaemoglobin standards has gone far to simplify evaluation of haemoglobinometry. Unfortunately, multi-channel instruments do not accept HiCN solutions directly and must be calibrated secondarily.

2. *Particles*

Universally accepted particle suspensions of known size distribution and concentration are not yet available, although a great variety of spores, pollen, fixed cells, and synthetic materials have been used (Lewis, 1972). Specialized instruments may provide a sufficiently accurate assessment of particle volume distribution against which threshold settings of clinical instruments can be judged (Thom, 1972).

3. *Survey validated material*

These materials are of unique value as the constituents have been analysed by large numbers of laboratories by many methods (see chapter 6).

B. OPERATIONAL VERSUS PHYSIOLOGICAL DEFINITION

Confusion may arise during the course of an evaluation if the operational characteristics of the kit or instrument are not understood.

Thus, for example, a RBC or WBC is, in reality, an estimate of the number of non-conductive particles above or between arbitrary electronic threshold limits, modified by statistical or empirical estimates of coincident passage, which have survived the collection, transportation, storage, and dilutional manipulations. This means that instruments may differ in their performance with different blood samples due to different thresholds (see p. 90).

With haemoglobin, too, measurement of absorbance of light at 540 nm after the addition of cyanide is no guarantee that the concentration of intra-erythrocytic Hb has been measured. The study of Gambino and Waraksa (1969) contrasting the HiCN method with the IL Haemoglobinometer is a good example of the operational modes of the instrument affecting the evaluation protocol. Because of that instrument's pipetting system, it was necessary to study the effects of whole blood viscosity.

Problems with measurement of PCV are discussed in chapter 8.

C. WHAT TO EVALUATE

1. *Manufacturer's claims*

One may attempt to establish if a manufacturer's claims are accurate or justified with respect to performance (i.e. precision and accuracy) and to operational or functional claims. At present in the U.S.A. this is the primary role of the Product Evaluation Sub-committee of the CAP; in Britain a similar function is performed by LDAG.

2. *Performance standards*

One may report on the performance or behaviour of an instrument, but such an evaluation may not be clinically useful unless compared to some performance which is based upon both state of the art, and medical usefulness criteria.

The FDA is empowered to set "product class standards" for *in vitro* diagnostic products, with the assistance of an advisory committee (Edwards, 1973). "A 'product class standard' is a statement describing performance requirements necessary to assure accuracy and reliability of results, specific labelling requirements necessary for the proper use of a particular class, and procedures for testing the product to assure its satisfactory performance."

3. *Comparative performance*

This type of evaluation is the most usual form and the Barnett–Youden scheme (1970) represents an experimental and statistical format for quantitative methods. Although a simpler system than the one under development by NCCLS, it is robust (see p. 21). The method under test is compared to a "reference" (or comparative method) and its performance is judged better or worse than the reference. Although designed for quantitative "kits" the protocol can be modified for the requirements of instrument systems.

4. *Engineering, costs, etc.*

During an evaluation, functional attributes of the method or instrument may intrude, often considerably. However, the evaluation is concerned primarily with performance and not with the engineering techniques which produce the end result. We do verify operational claims such as rapidity of test, training time, etc. The report to the manufacturer includes an impression of overall usability of the instrument.

D. MANDATORY OR VOLUNTARY PRODUCT EVALUATION

Most reports of product evaluation are non-governmental and independent. Some manufacturers approach recognized investigators and provide equipment and/or financial support in return for a product evaluation, while occasionally government agencies may contract for the evaluation of devices.

For the present, whether product evaluation is made mandatory by government decree or remains voluntary, there is not adequate organizational or laboratory facility to do an elaborate evaluation and periodic re-evaluation on the mass of available products. Much product evaluation (and a large part of the field engineering) is done at the expense of the pioneer purchasers. Much of the work done in the manufacturers' own laboratories is limited because access to adequately documented patient specimens is restricted and the rigours of a real clinical challenge are not met.

Fortunately, proficiency surveys can provide some degree of continuous user evaluation.

IV. Commonly Used Statistics in Product Evaluation

A. *F*-TEST OF PRECISION

This is based on the ratio of the variance of the test method divided by the variance of the reference method and allows estimates of the probability that one method is more precise than another. The *F*-test may be used to decide if a manufacturer's claimed precision ($CV\%$ or s) has been verified by assigning his claimed variance (s^2) an infinite degree of freedom (d.f.) and *F*-testing against s^2 observed during the experiment at $n - 1$ d.f.

Example:

Manufacturer claims $s = 2.5$, as a central tendency, Evaluator observes $s = 3.0$, $n = 20$ day to day tests. Is $s = 3.0$ really greater than $s = 2.5$? Place larger variance over smaller so $F > 1$. $(3.0)^2/(2.5)^2 = 1.44$. The critical value for F, d.f. 19/d.f.$_\infty$ $= 1.71$ ($P = 0.05$). Therefore $s = 3.0$ is not significantly greater than $s = 2.5$.

B. CONFIDENCE INTERVAL OF s

A simpler solution to the above problem is to consult a table of confidence intervals for s and determine if the 95% confidence interval of the observed s includes the claimed s (Natrella, 1966a).

C. STANDARD DEVIATION OF DIFFERENCES (s_d)

In a series of paired tests for a constituent, by two methods on the same specimen, computation of s of the *differences* between the paired values expresses the confidence interval in which future paired differences should lie. Approximately 95% of differences will be within $2s$ of each other. If the bias (systematic error) between the two methods is subtracted from each of the differences and the standard deviation calculated, the resulting s_d becomes an estimate of the proportional and random error between the two methods under the conditions of the experiment.

D. STANDARD DEVIATION OF DUPLICATES (s_{dup})

The standard deviation of differences between duplicate determinations on specimens is given by the formula $s_{dup} = \sqrt{\Sigma(d^2)/2n}$, where n = the number of duplicates.

In addition to its use as an estimate of precision it has clinical value as it gives an indication whether a change in a patient's test represents a real change or is within the expected imprecision interval.

E. PAIRED t-TEST

The significance of a difference between means (bias) of a series of paired values such as obtained in patient comparison studies is determined by the paired t-test, $t = |bias|\sqrt{n}/s_d$, using a two-tailed table of critical t-values, $P = 0.05$. Caution must be observed in interpreting the t-test, however, as poor precision or large random error between the methods obscures the significance of observed bias. Moreover, rigorous use of the t-test requires that the two population samples being compared have similar variance (s^2). This condition is usually fulfilled in the t-test grouped into pairs, but may not obtain when ungrouped data are used.

Even small biases becomes statistically significant when large numbers are used. The bias between the manual HiCN and the Coulter model S haemoglobin measurement (Table I) is based on thousands of values and the t-test is statistically significant, but one could hardly make a judgement as to clinical usefulness on that basis alone.

F. LINEAR REGRESSION

Analysis of line parameters is useful as most comparisons of methods involve the functional relationship: $Y = b_0 + b_1 x$.

Slope (b_1) is an estimate of proportional error between the methods and y-intercept (b_0) is an estimate of constant bias (error) between the methods similar to the bias of the t-test. Whereas the t-test bias is most accurate at the means of the two series, the y-intercept applies to the whole range of data, subject to goodness of fit of the data (confidence interval of line as a whole). The standard deviation of the differences between actual Y and Y_c calculated from the linear regression equation is similar to s_d, and is a measure of random error between the two methods.

Westgard and Hunt (1973) have contrasted the types of errors elucidated by t-test and linear regression (least squares) (Table IV).

TABLE IV.* Elucidation of errors by t-test or least squares

| | | Type of error | |
	Random	Constant	Proportional
Least squares			
slope	no	no	yes
y-intercept	no	yes	no
S_y	yes	no	no
t-test			
bias	no	yes	yes
s_d	yes	no	yes

* Modified from Westgard and Hunt (1973) by permission.

The least squares method of linear regression assumes that all imprecision of analysis occurs in Y and x is known precisely. In fact, unless the precision of method x (reference method) is substantially greater than Y, the method of group averages of Bartlett (1949) is preferred (Table V).

TABLE V. Comparison of t-test bias and s_d with y-intercept and S_y. Least squares versus group averages. Data: Phase chamber platelet counts (x) versus electronic counts (Y).

Platelet count $\times 10^9$/l	t-test bias	s_d	Linear regressions					
			Least squares			Group averages		
			y-int	S_y	Slope	y-int	S_y	Slope
0–50	3·09	4·80	2·32	4·98	1·03	2·34	4·78	1·03
50–150	6·75	21·65	−2·11	21·93	1·08	−10·77	21·64	1·17
150–450	−2·36	29·60	28·30	28·59	0·88	−9·16	30·24	1·02
450–850	−56·73	68·05	−10·90	70·62	0·93	−37·56	67·76	0·97

V. EVALUATION PROTOCOLS

A. NCCLS

Although still being extensively modified and tested, the conceptual approach of Westgard (personal communication) is worth reviewing. This protocol is based on the assumption that performance standards (PS) can be established for precision (day to day variation) and for accuracy (lack of bias between values obtained on patient specimens analysed by the method under evaluation and a well-accepted reference method). A decision tree determines if the data should be analysed by t-test parameters or by linear regression. If precision is acceptable then an estimate of the overall error is made by t-test and/or linear regression (least squares) statistics.

Briefly, the decision tree attempts to answer four questions by an elaborate series of calculations.

1. *Is precision of the test method acceptable?*

If twice the s (95% interval) of the day to day control pool values exceed *PS* in concentration units, the precision is not acceptable. Confirm by F-test.

2. *Should t-test or linear regression statistics be used?*

If the patient comparison data have an approximate 1:1 relationship t-test statistics may be used. If the range of numerical values is small, t-test statistics must be used.

The preferred statistics are those of linear regression which allows estimates of proportional error (slope), constant error (y-intercept) and random error between methods (S_y). However, reliable estimates require that the data have sufficient goodness of fit, and a suitably wide range of values.

3. *Is the maximum error between the methods acceptable?*

By t-statistics systematic and random error are summed, and if bias $+ 2s > PS$, the error is too large; if the t-test is positive at $P = 0.05$ the method is rejected.

By linear regression random and systematic errors are summed: if $|b_1 C + b_0)| - C + 2s > PS$, error is too large.

4. *Is the random variability (error) between methods acceptable?*

By t-test parameters if $2 s_d > 1.4(PS)$, random variability is too large. (This algorithm is derived from a simplifying assumption that in the patient comparisons each method contributes equally to the maximum acceptable variance, $(PS/2)^2$ for each, thus

$$2s_d = 2 \sqrt{\left(\frac{PS}{2}\right)^2 + \left(\frac{PS}{2}\right)^2} = 1 \cdot 4 \; PS.$$

For linear regression statistics if $2S_y > 1 \cdot 4(PS)$ the random variability is too large.

Since this protocol was devised substantial modifications have been made (Westgard *et al.*, 1974). Performance standard is defined as the allowable error at a concentration where critical medical decisions must be made. The different types of error are then isolated by experiment, quantitated by *t*-test and linear regression and finally summated.

(1) Random error is determined by the familiar day to day precision study.

(2) Proportional error is determined by recovery experiments.

(3) Constant error is determined by the addition of known interfering substances.

(4) Systematic error includes proportional and constant error and can be elucidated by linear regression of patient comparison studies and by calculation of the 95% interval for future mean values of Y (the W_2 of Natrella, 1966 *b*).

(5) Total error represents the grand summation of random and systematic error. Mathematically, this expression has two terms, the first representing average systematic error (linear regression at a critical medical decision level) and the second representing the uncertainty or imprecision of the method (the pooled sum of random error and W).

B. PES-C

A manufacturer desiring verification of claims submits statements as to precision and accuracy. Precision claims must be in statistical form such as coefficient of variation (CV) or *s*. Accuracy claims must be stated relative to other methods in a form that is experimentally verifiable.

1. *Precision*

Precision (reproducibility) analyses are performed once daily at two or three suitable concentration levels on stable control pools by the method under evaluation, and by a reference method. Twenty days are adequate. The levels selected ought to be at clinical decision levels, if possible. The reference method serves two purposes: (1) it acts as a control on the stability of the pools, and (2) it allows comparison of precision between the two methods.

2. *Accuracy*

Accuracy is determined relative to the reference method, and by recovery studies when applicable.

3. *Data plot*

A rectilinear plot of the patient comparison data provides visual evidence of linearity and displays outliers (see below). A useful procedure for methods with different units such as enzymes, is to draw lines for the normal limits perpendicular to the axis for each method and observe how many patients are considered normal or abnormal by the reference method and by the method under evaluation.

4. *Instrumental studies*

Tests for carryover and instrumental linearity are done. Studies to elucidate other aspects of instrumental behaviour are left to the discretion of the evaluator based on the claims the manufacturer hopes to verify.

5. *Statistical analysis*

Primarily F and t-test statistics are used although linear regression is considered useful, too.

6. *Verification*

All raw data, calculations and statistics are placed in the CAP archives. A written report is sent to the manufacturer and the chairman of the PES-C. The conclusions and supporting data are reviewed by the chairman of the CAP Standards Committee who issues the final letter of verification or non-verification of claims.

C. LABORATORY EQUIPMENT AND METHODS ADVISORY GROUP (U.K.)

The evaluation protocol used by Sharp and Ballard (1970) to evaluate the Coulter model S blends statistics and function. They include accuracy in normal and abnormal blood; precision; comparison with conventional methods; and such operational problems as costs, space, repairs and the "anxiety factor" which is quantitated as the "average hours/day spent fiddling with the contraption . . .".

The statistical description of the results are quite simple, consisting mostly of graphic displays of precision and linearity.

D. NON-QUANTITATIVE METHODS

The problems of precision and accuracy persist but are frequently more difficult to evaluate. For those methods which give categorical results in terms of "positive" or "negative", borderline results may be highly subjective. In such cases it may be necessary to include border-line plus–minus serum pools, (evaluated by a panel of analysts in a "blind" situation) to test the "precision" of the method.

Cichetti (1973) has devised a statistical method based on the central limit theorem for determining the accuracy of qualitative laboratory performance with a minimum of samples. This technique might also be useful in method comparisons, setting limits for the maximum number of disagreements permissible in a series of tests before the method under evaluation is considered invalid with respect to the reference. The comments of Feinstein (1973) on the handling of categorical data are also worth reading.

VI. Practical Problems in Product Evaluation

A. OBJECTIVITY

Regardless of the source of funding for the evaluation, the evaluator must impress upon the individuals performing the analyses the need for a dispassionate, detached attitude towards the results. Tests must be performed *exactly* as prescribed by the manufacturer in his *written* instructions. Departures or variations from the method should be documented and reported. The manufacturer should provide instruments, equipment, and reagents as ordinarily supplied to purchasers in regular containers, or they may be purchased through normal channels. Service and repairs should be those ordinarily performed for purchasers. An appropriate period of time for familiarization and calibration should preceed data collection.

1. *Number of analysts*

When possible, we prefer to have a single analyst perform all the tests.

2. *Number of instruments*

Ordinarily only a single instrument from the production line is tested. Should the original instrument be returned to the factory for major repairs because it does not function properly during initial studies a second instrument is used. Under these circumstances it is our personal preference to request return of the original instrument for comparative tests with the replacement.

3. *Number of evaluating laboratories*

As expense is a limiting factor, only one laboratory is assigned by the PES-C to do an evaluation.

B. SELECTION OF MATERIAL FOR PRECISION STUDY CONTROLS

The criteria are the same as those used for quality control. The material should resemble patient specimens; if it does not it may not have behaviour comparable to patient specimens. As a rough rule the CV of a replicate (within day) series on *patient* specimens should be about half the CV of the day to day pool, although with automatic counters they may be much closer. If day to day CV is *smaller* (such as occurs with a latex particle pool) then it can be taken that the pool does not behave as a patient specimen.

C. PATIENT COMPARISONS

1. *Number*

No evaluation is complete without extensive patient comparisons. Barnett and Youden (1973) stipulate 40, at no more than five per day. We have found that at least 100 comparisons are necessary to allow a more thorough evaluation of performance, especially if selected not only for high, low and normal values, but also for disease states. The larger number also permits the serendipitous discovery of less common interferences. These must be analysed over an extended period of time, rather than in a single day or week.

2. *Documentation*

Complete records should be available for each specimen analysed. This should include time of collection and of analysis, and have sufficient identification to permit review of the patient's chart, if necessary.

3. *Grouping of patient data*

In performing the statistical analysis it is useful to sub-group the patient comparisons in addition to comparing the whole set. A number of different sub-groups may be worth investigating.

(a) *By differences*. When the data are arranged in ascending pairs abrupt changes in the relative differences may be observed $(d/x_1 \times 100)$, indicating a change in the relative behaviour of the two methods.

(b) *By levels*. The data may be grouped into low, normal and high values to see if systematic and random error of the two methods changes with changes of concentration. If such changes are also observed in the precision data (day to day pool) at similar levels they support the validity of the pool as a replica of the test as performed on patient specimens.

(c) *By clinical levels*. Essentially similar to (b) but emphasizes clinical decision levels for disease entities.

(d) *By interfering factors*. Data may be grouped to test the effect of potential interferences. For example, Hb values may be grouped by white cell counts to ascertain the effects of leucocytes. From grouping by size distribution, we noted that specimens with very large platelets, >9 μm in diameter, were systematically undercounted by both electronic platelet counting methods we studied; and that a very slow erythrocyte sedimentation rate was also associated with severe undercounting, when platelet rich plasma was prepared by sedimentation as opposed to slow centrifugation. (Weisbrot and Ewing, 1974).

(e) *ANOVA*. This technique which is described in all standard statistics texts allows a more formal statistical analysis of the effects of grouping data by category.

D. SELECTION OF REFERENCE METHOD

Evaluations performed by the PES-C usually accept as reference the method designated as "well-accepted" on current CAP proficiency surveys.

1. *Haemoglobin*

The manual cyanmethaemoglobin method calibrated with certified HiCN standard is presently the method of choice. Although it has larger standard deviations on proficiency surveys than many automated methods, it is less subject to turbidity and other factors which cause inaccuracies.

2. *Particle counts*

The CAP specifies (at present) the Coulter model A, B, D or F as the well accepted method, in preference to the haemocytometer chamber, although the haemocytometer is a useful arbiter, and also permits morphological identification of the particles counted. Despite attacks on the *accuracy* of chamber counting (its imprecision is acknowledged) the data of Tables II and III illustrate its value. Sharp and Ballard (1970) used chamber counts for calibration and in our own laboratory we resort to them frequently.

3. *PCV*

The microhaematocrit remains the accepted method for CAP surveys, notwithstanding the elegance of the electronic methods. While the former may be affected by plasma trapping, excessive anticoagulant, and poikilocytosis (see chapter 8), errors may occur in electronic methods as the count may include large platelets and leucocytes in dyscrasias, as well as debris, particularly when the RBC is low.

E. ACCURACY

Arbitrarily, the product under test is considered accurate if the bias from the reference method is not significant by t-test. However, certain methods such as erythrocyte sedimentation do not yield to a statistical exercise alone. Bull and Brailsford (1972), in their evaluation of a rapid erythrocyte sedimentation method (Coulter Zetafuge), utilized medical significance data to test "accuracy". A significant bias may occur because the method under evaluation is more accurate, rather than less accurate, than the reference method. To prove this may be difficult in the absence of recovery studies. Brecher et al. (1956) utilized the expected stability of the MCV to solve this problem in their evaluation of the Coulter model A.

F. OUTLIERS

During any evaluation obvious discrepant results between the methods are often observed. Barnett and Youden permit one of the 40 patient comparisons to be arbitrarily discarded; if two or more are observed, all are included in the calculations. A simple plot of the data allows detection of potential outliers by inspection. We classify outliers into three categories.

1. *Mistakes*

Poor or spoiled specimens, errors in technique, instrument breakdown, mixups etc. are excluded without prejudice whenever identified, even if the values show good agreement.

2. *Functional (operational) outliers*

These are discrepant results ($>3s_d$) based on operational characteristics of the instruments (e.g. due to the differing effect of poikilocytes on the Coulter and microhaematocrits). This type of outlier can be detected from the specimen documentation data, or with a "secondary reference method" which is operationally similar to the method under evaluation. We perform statistics with and without such data. Final conclusion as to their importance requires judgement as to their clinical seriousness, frequency of occurrence, and ease of detection.

3. *True outliers*

If no cause for the discrepancy can be found, the value is considered a true outlier and it is retained in the data. We exclude none.

VII. Conclusions

Product evaluation has different implications for the manufacturer, the government, and the purchaser. But ultimately each wants to know what to expect from a product when installed at the end-point—the clinical laboratory. It is impossible for each laboratory to evaluate thoroughly each product prior to purchase and to select among competing diagnostic products wisely.

Eventually organizational unity needs to be established in the field of product evaluation, with accepted criteria for how an evaluation is to be done, and how and to whom the results should be made known.

REFERENCES

Barnett, R. N. and Youden, W. J. (1970). *Am. J. clin. Path.* **54,** 454–462.

Bartlett, M. S. (1949). *Biometrics* **5,** 207–212.

Brecher, G., Scheiderman, M. and Williams, G. Z. (1956). *Am. J. clin. Path.* **26,** 1439–1449.

Brittin, G. M. and Brecher, G. (1971). In "Progress in Hematology" (E. B. Brown and C. V. Moore, eds) vol. 7, pp. 299–341. Grune and Strattan, New York.

Bull, B. S. and Brailsford, J. D. (1972). *Blood* **40,** 550–559.

Canadian Standards Association (1973). "Clinical Laboratory and Electro-Medical Equipment Study". Rexdale, Ontario.

Cichetti, D. V., Keitges, P. and Barnett, R. N. (1974). *Health Lab. Sci.* **11,** 299–305.

Edwards, C. C. (1973). Federal Register, Vol. 38, Number 50, Part III, March 15.

Feinstein, A. R. (1973). *Clin. Pharm. Ther.* **14,** 898–915.

Gambino, S. R. and Waraksa, A. J. (1969). *Am. J. clin. Path.* **52,** 557–564.

Green, A. E., Middleton, V. L., Prentes, K. G. and Signy, A. G. (1969). *J. clin. Path.* **22,** 19–27.

Koepke, J. A. and Thakur, K. (1972). *Vox Sang.* **22,** 222–224.

Koepke, J. A. and Cichetti, D. V. (1973). *Transfusion* **13,** 41–43.

Lappin, T. R. J., Lamont, A. and Nelson, M. G. (1969). *J. clin. Path.* **22,** 11–18.

Lewis, S. M. (1972). In "Modern Concepts in Hematology" (G. Izak and S. M. Lewis, eds), pp. 217–229. Academic Press, New York and London.

Natrella, M. G. (1966a). "Experimental Statistics", chap. 2. National Bureau of Standards Handbook 91.

Natrella, M. G. (1966b). "Experimental Statistics", chap. 5. National Bureau of Standards Handbook 91.

Sharp, A. A. and Ballard, B. C. D. (1970). *J. clin. Path.* **23,** 327–335.

Thom, R. (1972). In "Modern Concepts in Hematology" (G. Izak and S. M. Lewis, eds) pp. 191–200 Academic Press, New York and London.

Weisbrot, I. M. and Ewing, N. S. (1974). *Am. J. clin. Path.* **62,** 693–701.

Westgard, J. O., Carey, R. N. and Wold, S. (1974). *Clin. Chem.* **20,** 825–833.

Westgard, J. O. and Hunt, M. R. (1973). *Clin. Chem.* **19,** 49–57.

16. Specimen Collection

L. S. SACKER

St Stephen's Hospital, London S.W.10, England

I. DEFINITION

The term "specimen collection" is a wide one even when limited to medical usage. It is, therefore, necessary to define the limitations which it is intended to apply to the term before commencing any discussions on the subject. For the present purposes the term will be confined to the collection by venous or cutaneous routes of samples of blood from the patient for haematological investigations. Collection also requires the use of a container into which the sample must be placed for transport to and for storage in the laboratory. Certain basic requirements are necessary for these containers: they must not leak the specimen, either when closed or on opening; and the material from which the container is manufactured must alter or damage the contents as little as possible in relation to the investigations to be performed.

Having defined the limits of this study it is pertinent to ask why a study of specimen collection is necessary. There are a number of reasons.

(1) With the increase in ease with which laboratory investigations can be performed and because of the increase in quality control which is coming to be applied at most stages, the sensitivity and accuracy of laboratory testing has been increased, such that even minimal variations in results can be detected which were not previously possible. It is, thus, inconsistent with scientific laboratory practice that properly controlled tests or investigations should be carried out on specimens which have been collected in an uncontrolled or haphazard way.

(2) Individual problems associated with specimen collection have been considered in the past, but this work has not been adequately analysed or repeated with newer techniques; nor have changing requirements been taken into account, and the methodology of specimen collections has remained unchanged and virtually unquestioned for many years.

II. Specimen Collection

The three sites from which blood is usually collected are: (1) the skin—capillary or arteriolar blood; (2) veins—venous blood; (3) arteries—arterial blood. As the collection of arterial blood is required for special purposes such as for blood gas analysis and requires different procedures and precautions, it is outside the scope of the present discussion.

A. CAPILLARY BLOOD

Capillary blood may be collected from the lobe of an ear or from a finger tip either from the pulp space area or from the lateral aspect over the terminal phalanx. In young infants blood is often collected from the plantar surface of the foot over the heel region.

There are difficulties in the collection of these specimens, which may lead to considerable variation, so that it may not be possible, for example, to obtain sequential haemoglobin estimations which agree with each other within an acceptable experimental error or which agree with the results obtained from venous blood taken at the same time (Dacie and Lewis, 1968). Better precision is possible if freely flowing drops of blood are obtained. Even then the flow of blood, and hence the reliability of the sampling, is influenced by the following factors.

1. *The temperature of the skin*

If the skin is cold or cyanosed capillaries will be closed and good blood flow will be difficult to achieve. It is best to warm the skin surface first, by rubbing the lobe of an ear until it is pink, or by warming the heel or hand in warm water.

2. *The depth of the incision*

Sterile prepacked disposable lancets are readily available and should be the only instruments used to make a suitable incision. They produce an incision of 2–3 mm in depth, which appears to be adequate at all the usual sites, unless the skin is particularly thickened. Lancets which are non-disposable will carry the risk of spread of infection, such as infective hepatitis. Lancets which require re-sterilization will tend to become blunted.

3. *Milking*

If a free flow of blood is not obtained there is a tendency to apply pressure by squeezing and while this may produce an adequate quantity of fluid the proportions between cellular and fluid components will be distorted (Fig. 1). If constriction is applied to the area to be sampled,

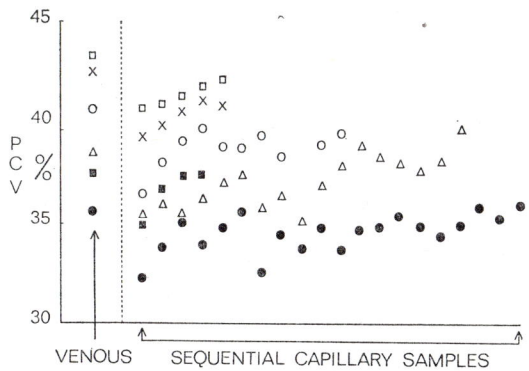

FIG. 1. Sequential PCV measurements on capillary blood from six normal donors compared with venous PCV taken at the same time. The initial capillary samples give low results which gradually approach the venous result. The probable explanation is the greater volume of interstitial fluid initially present.

such as a finger, the pressure must not be such as to produce cyanosis and must be easily relaxable to allow the arterial flow into the digit.

4. *Coagulation*

With the extensive tissue damage caused by the lancet incision thromboplastin is released and the coagulation process starts immediately. The time available for the collection of a specimen is therefore limited and platelet counts are unreliable as platelet aggregation occurs quickly.

The difficulties in obtaining accurate platelet counts from cutaneous samples were adequately appreciated by earlier pathologists. A multiplicity of techniques were evolved to try to overcome these problems (Wintrobe, 1942), and it is well recognized that it is undesirable to use capillary blood for platelet counts.

Difficulties affecting the counting of other cellular components have, however, not received similar attention. In brief, the use of capillary blood for haemoglobin or leucocyte estimations should be considered only if it is not possible to obtain suitable venous samples. Samples of capillary blood do, however, have their uses, particularly when it is not necessary to preserve the plasma:cell volume relationship in the sample. It is therefore reasonable to collect capillary blood for studies in which only red cells are used as, for example, in the sickle cell test or where the component to be measured is present equally in cells, plasma and intercellular fluid, as occurs with glucose or urea levels. It is reasonable to obtain larger volumes of clotted blood, usually from the heel in infants, for tests which require only serum.

Blood films made from capillary blood are preferable to those made from anticoagulated blood or from blood from a needle, but care must be taken that the blood is freely flowing and from a non-stagnant part to avoid alteration in the leucocyte distribution.

B. VENOUS BLOOD

The selected vein is usually in the ante-cubital fossa but veins in other parts of the arm can also be used. In young infants scalp veins or femoral veins may be the primary site of choice. Superficial veins such as are found in the forearms or arterio-sclerotic veins may be more difficult to canalize as they tend to slip away from the needle. Veins at the wrist are also more difficult to canalize as the overlying skin is firmer and they may be more painful when punctured.

Before canalizing a vein it is necessary to apply pressure by means of a sphygmomanometer or tourniquet and the pressure applied must be sufficient to impair the venous return but not to affect the arterial supply to the limb. A pressure of about 60 mm of mercury is usually adequate. Application of the tourniquet for a prolonged period of time causes loss of plasma from the blood held in the vein with a rise in the PCV, but an application of pressure for about a minute is necessary for this change to become significant (Mollison, 1972).

It is necessary to take precautions to prevent haemolysis when collecting the blood. The most potent cause of haemolysis is the introduction into the needle, and syringe if used, of spirit applied to cleanse the skin. It is advisable, after cleansing the skin with a suitable spirit-containing agent, such as isopropyl alcohol, either to allow adequate time for this spirit to evaporate to dryness or to wipe the skin with a dry sterile swab before inserting the needle. A spirit-containing swab should never be used over the puncture site after the needle has been removed as this may cause unnecessary pain and damage to the vein wall.

Posture affects the plasma proteins and haematocrit as fluid leaves the circulation on assuming the vertical posture after lying horizontal and it is desirable that posture should be standardized when a sample of venous blood is taken (Fawcett and Wynn, 1960; Tan *et al.*, 1973).

III. CHOICE OF NEEDLE

A wide range of needles are manufactured of differing gauge and length to meet the variable requirements for intradermal, intramuscular

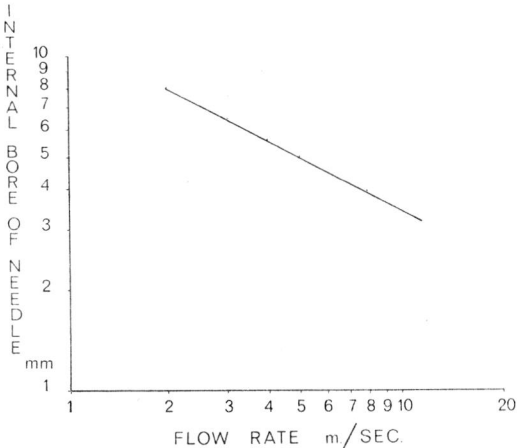

FIG. 2. Flow rates in needles of different internal bore size when blood is collected through them at a standard rate of 1 ml/s.

or intravenous injections as well as for venesection. Care must therefore be taken in the selection of the correct needle for venesection.

From the minimum internal bore of a variety of different needles of the same manufacturer (Table I), the calculated flow rates of a sample of blood collected at an arbitrary rate of 1 ml/s are shown in Fig. 2.

If these flow rates are to be obtained especially for the narrow gauge needles, this can only be done by the application of considerable suction pressure through the needle. The red-cell membrane can only withstand limited shearing stresses. A stress of the order of 4000 dynes/cm² will cause marked haemolysis and as some haemolysis can occur with much less shearing stress, there is a strong possibility of red cells being destroyed in the needle (Nevaril *et al.*, 1968).

If flow rates of this order are not obtained the time taken to collect the specimen may be unduly prolonged and coagulation is liable to interfere with the sample. The possibility of the plasma:cell ratio being upset is also possible with a slow laminar flow of blood in the needle (Macfarlane and Robb-Smith, 1961).

The choice of needle size has been further investigated using a sample reservoir system at a suitable height and of such a size as to produce a constant head of pressure of 100 mm of mercury. Different needles were inserted into the lower portion of a tube leading from this reservoir and the flow rates were measured. The experiment was carried out using a simple fluid such as water and a more complex fluid such as reconstituted plasma. The results obtained are presented in Table I. It can be seen from these results that the wider gauge needles have faster flow rates and this is entirely as predicted. Most shorter needles have faster flow rates than the longer needles of the same gauge but this does not apply so clearly to the narrower gauge needles. It would appear from these figures that the 25 mm 20 gauge needle is the smallest reasonable needle for the collection of blood from veins but the shorter 19 gauge and thin wall 19 gauge are better.

TABLE I. Needle variety and flow rate

Gauge	Needle Int. diam. (mm)	Length (mm)	Flow rates (ml/s) Water	Plasma	Change water/plasma (%)
19 TW	0·79	25	1·08	1·17	+8
19 TW	0·79	40	1·32	0·96	−27
19	0·65	40	1·10	0·75	−32
19	0·65	50	0·75	0·66	−12
20	0·56	25	0·64	0·56	−12·5
20	0·56	40	0·63	0·32	−49
21	0·5	40	0·39	0·29	−25·5
21	0·5	50	0·37	0·27	−27

19G Thin Wall needle is almost equivalent to 18G (internal diameter 0·8 mm). Other commonly used needles are 22G (0·4 mm) and 23G (0·32 mm); these have not been included in this study.

Of particular interest are the exceptions shown in Table I, where the longer 19 gauge thin wall needle gives a faster flow rate with water than the shorter version, and the 19 gauge 25 mm needle gives a faster flow rate with plasma than with water. There is also a variation in the reduction of the flow rate of plasma as compared with water for the different needle lengths. These changes may be due to design or manufacturing differences which can alter the flow pattern through the needle and further investigation is needed to sort out the optimal needle design.

The flow rate can be altered by applying suction pressure to the needle but it appears difficult to accept needles smaller than 20 gauge for venesection.

IV. COLLECTION SYSTEMS

There are two commonly used collection techniques: (1) the syringe and needle technique which may be modified to a needle and tube technique; (2) vacuum collection technique.

(1) *Syringe and needle.* A suitable needle is selected (see above) and attached to a syringe of appropriate size. After the syringe has been filled the needle is withdrawn from the vein and detached from the syringe. The pre-selected containers are then filled to their marks with blood and if an anticoagulant is present mixed gently.

The syringes used may be made either of glass or plastic. Plastic has the advantage of non-water-wettable surfaces, low individual purchase cost and if disposable requiring no cleaning or sterilizing. They do, however, create a problem of disposal and there is an increasing shortage of plastic which may make plastic syringes difficult to obtain. Glass syringes are expensive, require cleaning and sterilizing after each use and tend to have a high breakage rate.

(2) *Vacuum collection technique.* By this method a special two-ended needle is used which is fitted with a simple valve which will stop the flow of blood when a vacuum tube is not attached. This needle is fitted into a syringe-type barrel. When the needle is inserted into the vein a special container, with a vacuum which is designed to allow the entry of a predetermined volume of blood, is pierced by the other end of the needle, within the protection of the syringe barrel. When this container is filled it is replaced by further pre-selected containers which are filled in turn. By this means a number of samples of blood can be efficiently collected from one venepuncture and the individual containers should each be filled with the correct volume of blood. After use the needle is discarded and the syringe barrel reused, after sterilization if necessary.

Each procedure has its advantages and disadvantages:

(1) *Syringe technique.* Care must be taken when detaching the needle from the syringe and when filling the container that the inevitable aerosol or droplet fallout is kept to a minimum. Care must also be taken not to contaminate the outside of the containers when filling them, but this can be achieved without difficulty with proper training. It is also necessary to replace the caps to the containers firmly to prevent leakage.

The system permits great versatility in choice of container size and the material from which they are made. For some tests for which serum is required glass if preferable to plastic and for some coagulation studies plastic containers are essential.

(2) *Vacuum technique.* There appears to be less likelihood of aerosol and droplet contamination when collecting the specimen, but there is some evidence that this type of contamination exists in the presence of a spray of blood inside the syringe barrel and a droplet of blood on the bung of the vacuum container at the site of the needle puncture. There is no apparent risk in the outside of the container being more contaminated than this.

This has become a relatively versatile technique. A full range of containers of different sizes are available, but they are only glass as the presently available plastic materials are unsuitable. The main difficulty with vacuum containers occurs with their use in the laboratory. A simple bung extractor, which is readily available and cheap, causes unacceptable droplet and aerosol contamination and replacing a bung can also cause contamination of the outside of the container. Automatic aspiration devices are being developed but they will inevitably be expensive. Multiple sampling from one container is difficult, and the production of blood films is best achieved by separate skin prick techniques or from the sample of blood in the needle which may not be representative.*

There are also difficulties in the designs and production of vacuum type containers. These require a correct degree of vacuum for the sample of blood to be taken. The cap and sealing must be so designed that air cannot enter the vacuum container during storage so that the correct vacuum can be maintained. At present there is no standardization to control this vacuum and to ensure its maintenance until a stated possible expiry date. There is clearly an urgent need for this standardization to be undertaken.*

Mention should also be made of the labelling of the containers. This must be done at the patient's bedside to avoid mixing the specimens from different patients and to avoid delivery to the laboratory of

* These limitations are being overcome by design modifications.

unlabelled specimens. If the labelling is done before collection care must be taken that any labelled but unused containers are discarded rather than re-labelled for other patients as this is a common and dangerous source of confusion.

V. CONTAINERS

A. SPECIFICATION

A specimen container of known acceptable specifications is a fundamental requirement for safe laboratory usage and reliability. Such a specification has been published as BS 4851 by the British Standards Institution (1972), and a further extension of this work has been undertaken by the International Standards Organization, with the co-operation of the International Committee for Standardization in Hematology.

The aim of this specification is to encourage the development of specimen containers in which hazards are reduced to a minimum. Most hazards are due to the loss of contents and this is guarded against by tests for leakage and for spontaneous discharge which are described in BS 4851. The leak test makes reference to a torque value to which the caps are tightened so that no leakage will occur. It is quite possible to close every capped container to a torque value at which it will not leak but in a badly designed container this value may not be obtainable manually in the laboratory or may be too near the breaking point of the container. Table II gives the closure torque values of a range of

TABLE II. Leak tests closure torque values (torque values in N)

Supplier	0·113	0·266	0·339	0·452	0·565	0·678	0·791	0·904	1·017
1	+	−	−						
1	+	−	−						
2	+	−	−						
2	+	+	−						
2	+	+	−						
2	+	−	−		−			−	
2						+	++	−	−
3		+	−	−		−			
5		+	+	+		−			
1		−				−			
1	−	−	−						
1	+	−	−						
1	−	−		+					
7	+	−			−				
5					+	−	−	−	
3				+	−	−	−		

different containers made by the same or different suppliers. The best containers appear to require approximately equal torque values for closing and for opening.

A subsidiary advantage of standardization is to limit the number and variety of different containers available in the laboratory. It is necessary to stipulate maximum sizes to fit into commonly-used centrifuge buckets.

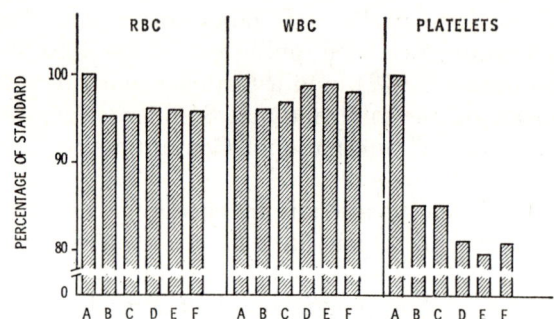

Fig. 3. Effect of specimen containers on blood count. The base (100%) was obtained from blood in siliconized glass. Reproduced from Lewis and Stoddart (1971) with permission. A = siliconized glass (standard); B = glass; C = polystyrene; D = polythylene, E = polycarbonate; F = polypropylene.

Another point dealt with in BS 4851 is the limit of interfering substances. The standard recommends the determination of elutable sodium and potassium as an index of contamination of the material from which the container is made. These are possible contaminants which are easy to estimate, but their absence clearly does not indicate freedom from contamination by other substances. To eliminate all interfering substances would involve a multiplicity of expensive tests which are not justified, provided that the container does not significantly alter the results of the specific investigation for which the contents are intended.

There is no clear evidence that the RBC, PCV or WBC is influenced by the container material although these results may be influenced by the anticoagulant used. Platelet counts are affected and platelets will adhere to the surfaces of both glass and plastic containers to a variable degree (Lewis and Stoddart, 1971). This has been demonstrated by scanning election microscopy using siliconized glass as a standard. It is shown that unsiliconized glass and polystyrene are equally effective in preventing adherence of platelets but are less effective than siliconized glass and that other plastics such as polyethylene, polycarbonate and polypropylene are marginally less effective still (Fig. 3).

The material from which the container has been made is of special importance when a good yield of serum is required, free of thrombin, such as for cross-matching investigations. With plastic containers clotting takes place slowly and clot retraction is poor. This is a direct effect of the surface because the clot retraction is inhibited by non-ionic surfaces with highly negative charges. Attempts have been made to overcome this deficiency of plastic by the use of additives or by gamma irradiation or glow discharge (Lewis and Stoddart, 1971). These methods are, however, expensive and reduce the shelf life of the material. The speeding up effect on the coagulation time is also variable, tends to be inadequate, and often produces haemolysis. The properties of unsiliconized glassware enable the coagulation process to proceed effectively to yield good serum samples.

The requirements for collecting blood samples for coagulation studies involve different needs. Glass contact activates the coagulation mechanism and the requirement is to use siliconized glass or plastic syringes and containers. The unsuitability of glass is not absolute as in citrated blood samples the prothrombin time appears to be unaffected for at least 4 h.

Another problem that requires further investigation is the erythrocyte sedimentation rate. The standard method for this test involves the use of thick walled glass tubing of stipulated bore size (ICSH, 1973). Current practice is increasingly towards the use of disposable equipment and the newer techniques involve the use of plastic tubes which are easier to manufacture with the correct bore size but the effect of this material on the accuracy of the test has not been fully elucidated and more investigation is needed.

With some techniques, such as in the use of heat-damaged red cells (Corner et al., 1970) or in grouping and cross-matching reactions the results may be influenced by the heat conductivity of the material from which the container is made. The results of heat conductivity studies are shown in Figs 4 and 5.

The essential conclusions are that no single material for containers will satisfy all needs and that all container materials should be examined for their effects on the planned investigations.

B. USE IN THE LABORATORY

Containers designed for the collection of specimens are also used for blood storage within the laboratory. The main difficulty in this use is caused by the need for frequent sampling from the container. With each additional opening there are increased risks of forcing some of the contents between the cap and screw thread and contaminating the outside of the container. With vacuum containers the same problems

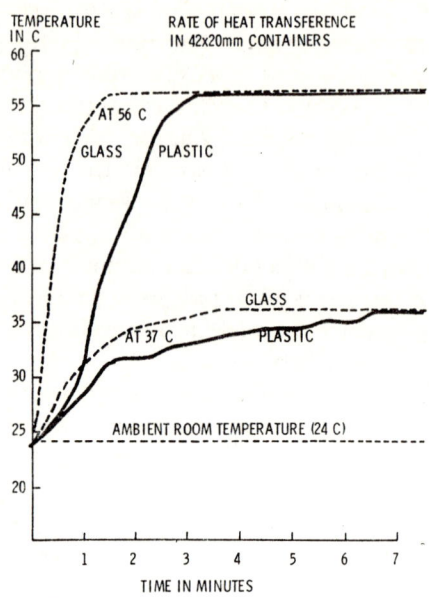

Fig. 4. Heat transference in glass and plastic containers. From the late Dr S. Wray.

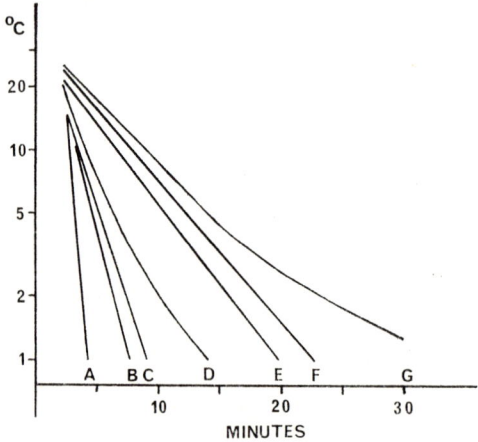

Fig. 5. Gradient fall of temperature on cooling. A, C and D are glass containers (3, 16 and 25 ml respectively): B, E, F and G are polystyrene containers (3, 12, 16 and 25 ml respectively). Reproduced from N. Corner, S. Ahuja and S. M. Lewis (1970) with permission.

exist, as replacement of the bung forces some of the contents on to the outside of the tube.

In my laboratory screw capped containers are used and are opened between two and four times for sampling. The first opening is to make a blood film and the second is to offer the sample to an automated cell counter. Prior to both these openings the container is on a mixing machine and the contents are in contact with the cap. Later openings are required for additional investigations or any repeat tests. In practice few of our containers are contaminated on the outside after two openings. Containers which are fitted with bungs are more difficult to open without aerosol or droplet contamination but if the blood film is made at the time the specimen is taken, either from a drop of blood in the needle or by separate skin puncture, then it may only be necessary to open these containers once, as a routine, and it may be possible with the further development of automated apparatus to aspirate contents from the container which need not be opened except for additional investigations.

VI. ANTICOAGULANTS

Haematological tests frequently require an additive to prevent the blood clotting. The additive or anticoagulant used depends on the test for which the container is designed. The commonly used anticoagulant for cell counting is one of the salts of ethyelne diamine tetraacetic acid (EDTA) (Dacie and Lewis, 1968). This has replaced ammonium and potassium oxalate mixture as the routine anticoagulant for cell counting because with the latter platelet clumping occurs, thus preventing platelet counting, and furthermore the anticoagulant effect on the white cells is more severe than with EDTA. Heparin is also not used for cell counting because of its clumping effect on platelets and leucocytes and because blood films made from it are difficult to stain clearly. Heparin is however invaluable for obtaining unhaemolysed and unaltered red cells for fragility studies and for red-cell enzyme estimations.

Trisodium citrate is the usual anticoagulant for coagulation investigations and for the erythrocyte sedimentation rate.

Only EDTA and trisodium citrate anticoagulants will be considered further in the context of this article.

A. ETHYLENE DIAMINE TETRA-ACETIC ACID

A major advantage of EDTA is that it is an efficient anticoagulant which does not affect the cell size or numbers, and it is effective in platelet preservation for platelet counting.

Ethylene diamine tetra-acetic acid acts as a chelating agent. The molecule of EDTA has four hydrogen atoms in its formula which have normal valency linkages (Fig. 6), and it is by substitution at these points that other elements are attached to form salts of EDTA. One molecule of EDTA or its salts can chelate one molecule of calcium, and the attachment is via four "co-ordinate" bonds on the molecule of EDTA which react and form a union with four donor bonds on the calcium molecule. The union is firm and the calcium ion can only be replaced by another substance which has greater affinity for the chelating agent than the calcium.

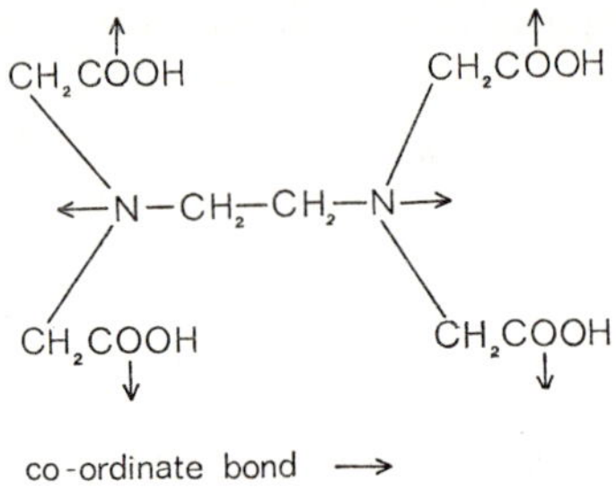

$$CH_2COOH \qquad CH_2COOH$$

$$\leftarrow N-CH_2-CH_2-N\rightarrow$$

$$CH_2COOH \qquad CH_2COOH$$

co-ordinate bond $\longrightarrow$

Fig. 6. Formula of ethylenediaminetetra-acetic acid (EDTA), mol. wt 292.

As EDTA itself is not easily soluble it is not a satisfactory substance to use as an anticoagulant. The solubility can be increased by forming salts by normal valency linkage substitution at the COOH positions. Because the cations at these positions ionize, the solubility is increased, but the presence of a cation at these positions has no effect on the chelating power of the molecule. The usual cations to replace the hydrogen at these positions are lithium, sodium or potassium and any number of the four hydrogen atoms may be replaced (Sacker et al., 1959). In practice the commonly used anticoagulants are the disodium or dipotassium salts of EDTA. The latter is slightly more soluble, and therefore has tended to be more commonly used. The tripotassium salt is also used for the same purpose and there is some evidence that this is a better chelating agent than the dipotassium salt.

A difficult question to answer is what is the optimal amount of EDTA needed to provide efficient anticoagulation. Based on the acid 0·44 mg of EDTA is needed to chelate the calcium in 1 ml of whole blood, assuming a concentration of 6 mg%. In practice more is used than this as allowance has to be made for the higher calcium content of anemic blood. The actual amount of EDTA used is really a compromise between the amount required to produce efficient anticoagulation and the amount required to produce minimal cellular damage. In BS 4851 the recommendation is to use 1·5 mg ± 0·25 of dipotassium EDTA per ml of blood. This recommendation relates only to dipotassium EDTA. As other salts, especially the tripotassium salt, should not be excluded, the chelating power should be related to EDTA itself, and for this purpose the equivalent is approximately 1·2 mg of EDTA per ml of blood. To produce anticoagulation a figure of about 0·75 mg EDTA per ml is a reasonable lower limit.

The physical form in which the EDTA is put into the container is of importance. Hard masses of amorphous material tend to be insoluble and some manufacturers have over-compensated for this by increasing the amount of EDTA up to 2–2·5 mg per ml. This may reduce the number of clotted specimens but it renders cell morphology useless. Higher concentrations affect other parameters as well (see below). Some manufacturers overcome this difficulty by spraying the correct amount of EDTA salt in solution into the container or rotating the container whilst drying.

The concentration of EDTA does not appear to affect the Hb, RBC, PCV, MCV or total leucocyte count unless the level of EDTA is increased to 4 mg per ml. This concentration may be reached if insufficient blood is taken into the container. Prolonged contact time of more than 24 h produces similar changes. The changes in cell morphology commence within half an hour of contact and these changes can affect not only leucocytes but also red cells and platelets. The type of change noticed in the leucocytes is variable in that in an individual's blood, cellular changes vary in intensity and affect progressively increasing numbers of cells. There is also some degree of resistance to change in some individual's leucocytes which gives rise to a variation in the amount of anticoagulant change, especially with lower concentrations of EDTA.

The first changes in the polymorphs are swelling and loss of structure in the polymorph lobes; stretching of the interlobular bridges then occurs with loss of granulation in the cytoplasm. Vacuolation occurs in the nucleus or cytoplasm of the cells. These changes progress with more marked nuclear swelling and there then appears to be a cross-over of the nuclear chromatin, giving rise, in its more extreme forms to the typical clover leaf pattern. Ultimately this change progresses to complete morphological disintegration of the cell, leaving a bare nucleus which can still be counted in a particle counter. These changes are repesented in Fig. 7. Similar changes occur in the mononuclear cells.

Platelets appear to swell, leaving giant platelets which then disintegrate causing an artificially high platelet count as these fragments are large enough to be counted as morphologically normal platelets.

POLYMORPH MORPHOLOGY

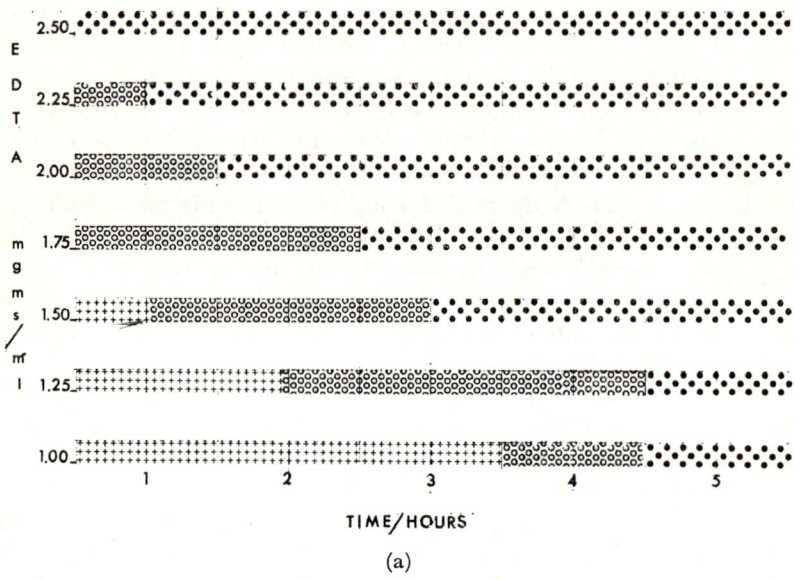

(a)

EFFECT OF EDTA ON CELL MORPHOLOGY

NEUTROPHILS

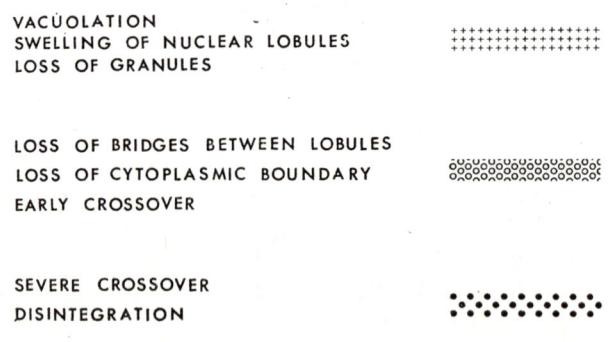

(b)

Fig. 7. Effect of concentration of EDTA and time of contact on polymorph morphology.

Red cells also change in shape with increasing anisocytosis and spherocytosis, and later macrocytosis which is accompanied by an increasing mean cell volume as the time of contact with EDTA increases. Ultimately some of these cells are destroyed.

B. TRISODIUM CITRATE

This anticoagulant, at a concentration of 31 g/l of $Na_3C_6H_5O_7.2H_2O$, is used for two main purposes: (1) for coagulation investigations it is used as a fluid anticoagulant, one volume of the trisodium citrate solution to nine volumes of blood; (2) for erythrocyte sedimentation rate one volume of the anticoagulant is mixed with four volumes of blood.

The importance of a correct procedure of specimen collection is well illustrated, for example, with the ESR. Even when this test is carried out under standard conditions as recommended by ICSH (1973), faulty specimen collection may invalidate the result. When screw-capped containers are used, if the cap becomes loosened during storage evaporation or spillage of the contents occurs with resulting inaccuracy in the test. Thus, consideration needs to be given to the use of a solid anticoagulant which can be dried in the container.

In a simple experiment comparison was made in 30 tests, between the standard ESR technique, and ESR on blood collected into (a) dried trisodium citrate, and (b) dried trisodium citrate to which saline was added immediately before the test, to bring the total volume and proportions up to those used in the standard technique.

The result was that in all but one case there was an increased ESR with dried citrate as compared with the standard technique, and the diluted solid citrate technique gave lower readings in 11 out of 30 as compared with the standard technique.

There are theoretical advantages in using solid anticoagulants but more work is required to determine the standardization of the ESR test using dried citrates. Similarly, a solid anticoagulant for coagulation investigations might have considerable advantage, but this might necessitate new standards for procedures and interpretation of results of tests.

VII. TRANSPORT

It is necessary, when packing specimens for transport, that adequate precautions are taken to prevent the container being damaged in transit and to provide sufficient protective wrapping round the container to hold the contents should fracture or leakage occur. For simple inter-hospital transport placing the specimens in sealed plastic bags in boxes is adequate. For long distance transport or postal transmission more efficient protective wrapping is necessary and in most areas there will be local post office rules which will be applicable. It is recommended that the specimen container should be enclosed in a despatch shell which may be made of wood, hardboard or plastic and must be able to withstand considerable outside pressure without distortion and must hold the contents of the specimen containers if fractured without leakage. This despatch shell is then placed in a bag or box for transmission and delivery.

Containers, especially the screw-thread type, have often been found to leak in transit. There may be a number of reasons for this. The cap of the container may not have been adequately tightened originally. This defect is surprisingly difficult to overcome as it involves the human element. If tightened adequately the cap may still be loosened by vibration during transit, by temperature change with different expansion characteristics between the different materials of the container and its cap or by plastic "creep" between different plastic surfaces.

The containers also need to be protected from extremes of temperature when being transported and from variation in pressure as may occur when being transported in unpressurized compartments of aircraft. Reference has been made in BS 4851 to a requirement for specimen containers being transmitted by tubular conveyor systems. These systems tend to be little used because of the risk to the specimen but safer containers may encourage their use.

There is also a requirement for standardization of the procedure when transmitting specimens across national boundaries. Specimens often deteriorate due to unnecessary delay at customs points and inadequate storage at the wrong temperature whilst awaiting or during transport despite adequate instructions on the labels. As the transmission of samples under these circumstances is increasing, the requirement for an internationally agreed procedure is urgent.

VIII. Conclusion

This chapter on specimen collection gives an account of some of the problems involved and outlines some experimental work which questions predetermined concepts. Clearly, the reliability of haematological tests will be affected by the method of blood collection, by the design and material of the container into which the blood is collected and by the anticoagulant which is used. These, and other aspects of specimen collection should be subject to standardization and quality control no less stringent than for the tests themselves.

ACKNOWLEDGEMENT

It was originally intended that this article would be written jointly with Dr Stanley Wray. Unfortunately owing to Dr Wray's sudden death this was not possible. Much of the article, especially the sections on containers, is based on his work.

REFERENCES

British Standards Institution (1972). "Specification for Medical Specimen Containers for Haematology and Biochemistry, BS 4851".

Corner, N., Ahuja, S. and Lewis, S. M. (1970). *J. clin. Path.* **23**, 646.

Dacie, J. V. and Lewis, S. M. (1968). "Practical Haematology", 4th edn, pp. 9–11. Churchill, London.

Fawcett, J. K. and Wynn, V. (1960). *J. clin. Path.* **13**, 304–310.

International Committee for Standardization in Hematology (1973). *Br. J. Haemat.* **24**, 671–673.

Lewis, S. M. and Stoddart, C. T. H. (1971). *Lab. Pract.* **20**, 787–792.

Macfarlane, R. G. and Robb-Smith, A. H. T. (1961). "Functions of the Blood", pp. 303–305. Blackwell, Oxford.

Mollison, P. L. (1972). "Blood Transfusion in Clinical Medicine", 5th edn, pp. 112–113. Blackwell, Oxford.

Nevaril, C. G., Lynch, E. C., Alfrey, C. P. and Hellums, S. D. (1968). *J. Lab. clin. Med.* **71**, 784–790.

Sacker, L. S., Saunders, H. E., Page, B. and Goodfellow, M. (1959). *J. clin. Path.* **12**, 254–257.

Tan, M. H., Wilmhurst, E. G., Gleason, R. E. and Soeldner, J. S. (1973). *New Engl. J. Med.* **289**, 416–418.

Wintrobe, M. (1942). "Clinical Haematology", p. 137. Kimpton, London.

17. The Cost of Quality Control

J. F. COSTER

National Institute of Public Health, Bilthoven, The Netherlands

It is fairly easy to determine the total cost of running a laboratory, by adding together the cost of accommodation, labour and materials. It is less easy to calculate the cost per test, as this either requires an assumption that each test has equal weight in terms of labour and materials, or the relative load for each test (i.e. test units) must be assessed. Dividing total cost by test units gives a cost per test, and this system has been established in several countries for working out fees or remuneration. However, it was soon discovered that this cost per test is not the same in all laboratories, and any one test might be costed many times greater in one laboratory than in another.

The first question to be considered is whether the most expensive is the best. If this were the case, the problem of determining the expense of quality control would be solved, as the cost of quality would be the difference between the most expensive and the cheapest test cost. Unfortunately, this state does not hold true. What criteria then can be used as the indicator of the cost for quality? The basic principles of quality control and their cost have been described by Eilers (see chapter 1). It should be emphasized, however, that in practice the assessment is complex and that it is necessary to distinguish between the expense of introducing quality control when a new laboratory is set up and the cost of trying to get an "established" laboratory in line.

When a new laboratory is founded, the director should determine what technique and equipment are most appropriate for each test that will be undertaken, the type of personnel required and what controls to introduce. If he is wise in these selections, the only costs of control will be the purchase of control samples, the cost per test (see above) for repeat testing of selected samples from the routine specimen turnover and the expenses for statistical analysis of data.

A completely different problem is present in the case of an existing laboratory, where a routine procedure of operation has become an established habit, and where suddenly, either as a result of an inter-laboratory survey or because of disbelief on the part of the clinicians, some tests fall into disrepute. Many laboratories have been in this state at one time or another and experience has shown that the extent of work and the cost which must be borne in order to get acceptable results varies enormously, depending on the test or series of tests under consideration. In Fig. 1 an attempt is made to give a cost-improvement

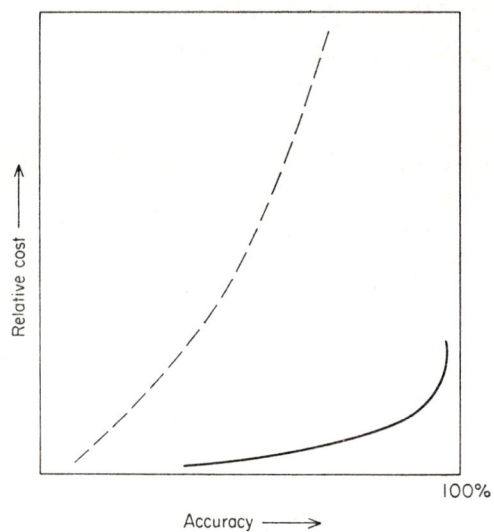

FIG. 1. Cost-improvement ratio for laboratory tests. The continuous line shows the type of reaction where a light input of effort and money bring the accuracy to the desired value near 100%; the dotted line shows a reaction which is more difficult to control, where every input in cost (e.g. better equipment or personnel) gives only a slight improvement without coming as near to the goal as desired. In reality, the lines should be drawn step-wise.

ratio for different circumstances. Any laboratory can, with a minimum of endeavour and cost, get good results in a haemoglobinometry trial. However, even the "best" laboratories must strive hard to attain necessary improvement with some of the more complex tests. The "established" laboratory may have to overcome the problem of having less able personnel performing the test. To teach poor workers may prove more expensive than appointing better qualified people at once. Another problem is caused by the laboratory having equipment which is inferior, but not yet fully obsolete, necessitating the abandonment of equipment which cannot be written off routinely as a simple administrative procedure.

What conclusion can be drawn on the cost of quality and quality control? From the foregoing, it will be evident that the percentage of the laboratory cost devoted to that purpose will differ in each individual laboratory. A range as wide as 15–40% of a laboratory cost has been suggested as necessary to ensure satisfactory quality management. Some published surveys and reports by official authorities in various countries have indicated a fixed percentage as a median value. Such values can be discounted as unreal, unless all of the associated cost has been included. Moreover, a meticulously operated laboratory with good direction and higher paid skilled workers needs only scanty control, whereas a much higher percentage of the laboratory budget must be devoted to quality control if the laboratory itself is less well provided with professional expertise. Beyond doubt, adequately paid well-trained staff of high calibre provides the basis for accurate and efficient work which is the essential feature of a reliable laboratory.

SUBJECT INDEX